Methods in Enzymology

Volume 344
G PROTEIN PATHWAYS
Part B
G Proteins and Their Regulators

METHODS IN ENZYMOLOGY

EDITORS-IN-CHIEF

John N. Abelson   Melvin I. Simon

DIVISION OF BIOLOGY
CALIFORNIA INSTITUTE OF TECHNOLOGY
PASADENA, CALIFORNIA

FOUNDING EDITORS

Sidney P. Colowick and Nathan O. Kaplan

Health Reference Series

Adolescent Health Sourcebook
AIDS Sourcebook, 1st Edition
AIDS Sourcebook, 2nd Edition
AIDS Sourcebook, 3rd Edition
Alcoholism Sourcebook
Allergies Sourcebook, 1st Edition
Allergies Sourcebook, 2nd Edition
Alternative Medicine Sourcebook, 1st Edition
Alternative Medicine Sourcebook, 2nd Edition
Alzheimer's, Stroke & 29 Other Neurological Disorders Sourcebook, 1st Edition
Alzheimer's Disease Sourcebook, 2nd Edition
Arthritis Sourcebook
Asthma Sourcebook
Attention Deficit Disorder Sourcebook
Back & Neck Disorders Sourcebook
Blood & Circulatory Disorders Sourcebook
Brain Disorders Sourcebook
Breast Cancer Sourcebook
Breastfeeding Sourcebook
Burns Sourcebook
Cancer Sourcebook, 1st Edition
Cancer Sourcebook (New), 2nd Edition
Cancer Sourcebook, 3rd Edition
Cancer Sourcebook for Women, 1st Edition
Cancer Sourcebook for Women, 2nd Edition
Cardiovascular Diseases & Disorders Sourcebook, 1st Edition
Caregiving Sourcebook
Childhood Diseases & Disorders Sourcebook
Colds, Flu & Other Common Ailments Sourcebook
Communication Disorders Sourcebook
Congenital Disorders Sourcebook
Consumer Issues in Health Care Sourcebook
Contagious & Non-Contagious Infectious Diseases Sourcebook
Death & Dying Sourcebook
Depression Sourcebook
Diabetes Sourcebook, 1st Edition
Diabetes Sourcebook, 2nd Edition
Diabetes Sourcebook, 3rd Edition
Diet & Nutrition Sourcebook, 1st Edition
Diet & Nutrition Sourcebook, 2nd Edition
Digestive Diseases & Disorder Sourcebook
Disabilities Sourcebook
Domestic Violence & Child Abuse Sourcebook
Drug Abuse Sourcebook
Ear, Nose & Throat Disorders Sourcebook
Eating Disorders Sourcebook
Emergency Medical Services Sourcebook
Endocrine & Metabolic Disorders Sourcebook
Environmentally Induced Disorders Sourcebook
Ethnic Diseases Sourcebook
Eye Care Sourcebook, 2nd Edition
Family Planning Sourcebook
Fitness & Exercise Sourcebook, 1st Edition
Fitness & Exercise Sourcebook, 2nd Edition
Food & Animal Borne Diseases Sourcebook
Food Safety Sourcebook
Forensic Medicine Sourcebook
Gastrointestinal Diseases & Disorders Sourcebook

Teen Health Series
Helping Young Adults Understand, Manage, and Avoid Serious Illness

Diet Information for Teens
Health Tips about Diet and Nutrition
Including Facts about Nutrients, Dietary Guidelines, Breakfasts, School Lunches, Snacks, Party Food, Weight Control, Eating Disorders, and More

Edited by Karen Bellenir. 399 pages. 2001. 0-7808-0441-4. $58.

"Full of helpful insights and facts throughout the book. . . . An excellent resource to be placed in public libraries or even in personal collections."
—*American Reference Books Annual 2002*

"Recommended for middle and high school libraries and media centers as well as academic libraries that educate future teachers of teenagers. It is also a suitable addition to health science libraries that serve patrons who are interested in teen health promotion and education." —*E-Streams, Oct '01*

"This comprehensive book would be beneficial to collections that need information about nutrition, dietary guidelines, meal planning, and weight control. . . . This reference is so easy to use that its purchase is recommended." —*The Book Report, Sep-Oct '01*

"This book is written in an easy to understand format describing issues that many teens face every day, and then provides thoughtful explanations so that teens can make informed decisions. This is an interesting book that provides important facts and information for today's teens." —*Doody's Health Sciences Book Review Journal, Jul-Aug '01*

"A comprehensive compendium of diet and nutrition. The information is presented in a straightforward, plain-spoken manner. This title will be useful to those working on reports on a variety of topics, as well as to general readers concerned about their dietary health."
—*School Library Journal, Jun '01*

Drug Information for Teens
Health Tips about the Physical and Mental Effects of Substance Abuse
Including Facts about Alcohol, Anabolic Steroids, Club Drugs, Cocaine, Depressants, Hallucinogens, Herbal Products, Inhalants, Marijuana, Narcotics, Stimulants, Tobacco, and More

Edited by Karen Bellenir. 452 pages. 2002. 0-7808-0444-9. $58.

"This is an excellent resource for teens and their parents. Education about drugs and substances is key to discouraging teen drug abuse and this book provides this much needed information in a way that is interesting and factual." —*Doody's Review Service, Dec '02*

Mental Health Information for Teens
Health Tips about Mental Health and Mental Illness
Including Facts about Anxiety, Depression, Suicide, Eating Disorders, Obsessive-Compulsive Disorders, Panic Attacks, Phobias, Schizophrenia, and More

Edited by Karen Bellenir. 406 pages. 2001. 0-7808-0442-2. $58.

"In both language and approach, this user-friendly entry in the *Teen Health Series* is on target for teens needing information on mental health concerns." —*Booklist, American Library Association, Jan '02*

"Readers will find the material accessible and informative, with the shaded notes, facts, and embedded glossary insets adding appropriately to the already interesting and succinct presentation."
—*School Library Journal, Jan '02*

"This title is highly recommended for any library that serves adolescents and parents/caregivers of adolescents." —*E-Streams, Jan '02*

"Recommended for high school libraries and young adult collections in public libraries. Both health professionals and teenagers will find this book useful."
—*American Reference Books Annual 2002*

"This is a nice book written to enlighten the society, primarily teenagers, about common teen mental health issues. It is highly recommended to teachers and parents as well as adolescents."
—*Doody's Review Service, Dec '01*

Sexual Health Information for Teens
Health Tips about Sexual Development, Human Reproduction, and Sexually Transmitted Diseases
Including Facts about Puberty, Reproductive Health, Chlamydia, Human Papillomavirus, Pelvic Inflammatory Disease, Herpes, AIDS, Contraception, Pregnancy, and More

Edited by Deborah A. Stanley. 400 pages. 2003. 0-7808-0445-7. $58.

"As a reference for the general public, this would be useful in any library." —*E-Streams, Jun '01*

"Provides helpful information for primary care physicians and other caregivers interested in occupational medicine.... General readers; professionals."
—*Choice, Association of College & Research Libraries, May '01*

"Recommended reference source."
—*Booklist, American Library Association, Feb '01*

"Highly recommended." —*The Bookwatch, Jan '01*

Worldwide Health Sourcebook
Basic Information about Global Health Issues, Including Malnutrition, Reproductive Health, Disease Dispersion and Prevention, Emerging Diseases, Risky Health Behaviors, and the Leading Causes of Death

Along with Global Health Concerns for Children, Women, and the Elderly, Mental Health Issues, Research and Technology Advancements, and Economic, Environmental, and Political Health Implications, a Glossary, and a Resource Listing for Additional Help and Information

Edited by Joyce Brennfleck Shannon. 614 pages. 2001. 0-7808-0330-2. $78.

"Named an Outstanding Academic Title."
—*Choice, Association of College & Research Libraries, Jan '02*

"Yet another handy but also unique compilation in the extensive Health Reference Series, this is a useful work because many of the international publications reprinted or excerpted are not readily available. Highly recommended." —*Choice, Association of College & Research Libraries, Nov '01*

"Recommended reference source."
—*Booklist, American Library Association, Oct '01*

8. Bosma TJ, Kennedy MA, Bodger MP, Hollings PE, Fitzgerald PH: Basophils exhibit rearrangement of the bcr-gene in Philadelphia Chromosome positive chronic myeloid leukemia. Leukemia 1988;2:141-48.
9. Butterfield JH, Weiler D, Dewald G, Gleich GJ: Establishment of an immature mast cell line from a patient suffering from mast cell leukemia. Leuk Res 1988;12:345-50.
10. Columbo M, Horowitz EM, Botana LM, et al. The human recombinant c-kit receptor ligand, rhSCF, induces mediator release from human cutaneous mast cells and enhances IgE dependent mediator release from both skin mast cells and peripheral blood basophils. J Immunol 1992;149:599-608.
11. Denburg JA, Abrams J, HowieK, Girgis-Garbarro A, Harnish D. Interleukin-5 is a human basophilopoietin. Blood 1991;771:462-8.
12. Denburg JA, Telizyn S, Messner H, Lim B, Jamal N, Ackerman SJ, Gleich JG, Bienenstock J. Heterogeneity of human peripheral blood eosinophil-type colonies: Evidence for a common basophil- eosinophil progenitor. Blood 1985;66:312-8.
13. Denburg JA, Wilson WEC, Bienenstock J: Basophil production in myeloproliferative disorders: increases during acute blastic transformation of chronic myeloid leukemia. Blood 1982;60:113-9.
14. Flanagan JG, Chan DC, Leder P. Transmembrane form of the kit ligand growth factor is determined by alternative splicing and is missing in the Sld mutant. Cell 1991;64:1025-35.
15. Furitsu T, Saito H, Dvorak AM, et al. Development of human mast cells in vitro. Proc Natl Acad Sci (USA) 1989;86:10039-43.
16. Furitsu T, Tsujimura T, Tono T, et al. Identification of mutations in the coding sequence of the proto-oncogene c-kit in a human mast cell leukemia cell line causing ligand-independent activation of the c-kit product. J Clin Invest 1993;92:1736-44.
17. Galli SJ. Biology of disease. New insights into 'The riddle of the mast cells': Microenvironmental regulation of mast cell development and phenotypic heterogeneity. Lab Invest 1990;62:5-33.
18. Ganser A, Lindemann A, Seipelt G, Ottmann OG, Herrmann F, Eder M, Frisch J, Schulz G, Mertelsmann R, Hoelzer D. Effects of recombinant human interleukin-3 in patients with normal hematopoiesis and in patients with bone marrow failure. Blood 1990;76:666-71.
19. Ghildyal N, McNeil P, Gurish MF, Austen KF, Stevens RL. Transcriptional regulation of the mucosal mast cell-specific protease

gene, MMCP-2, by interleukin-10 and interleukin-3. J Biol Chem 1992;8473-7.
20. Hirai K, Morita YM, Misaki Y. et al. Modulation of human basophil histamine release by hemopoietic growth factors. J Immunol 1988;141:3958-64.
21. Ishizaka T, Dvorak AM, Conrad DH, Niebyl JR, Marquette JP, Ishizaka K. Morphologic and immunologic characterization of human basophilis developed in cultures of cord blood mononuclear cells. J Immunol 1985;134:532-40.
22. Ishizaka T, Ishizaka K. Activation of mast cells for mediator release through IgE receptors. Prog Allergy 1984;34:188-235.
23. Kirshenbaum AS, Goff JP, Kessler SJ, Mican JM, Zsebo KM, Metcalfe DD: Effect of IL-3 and stem cell factor on the appearance of human basophils and mast cells from $CD34^+$ pluripotent progenitor cells. J Immunol 1992;148:772-783.
24. Kirshenbaum AS, Kessler SW, Goff JP, Metcalfe DD: Demonstration of the origin of human mast cells from $CD34^+$ bone marrow progenitor cells. J Immunol 1991;146:1410-5
25. Kitamura Y, Yokoyama M, Matsuda H, Ohno T, Mori KJ. Spleen colony forming cell as common precursor for tissue mast cells and granulocytes. Nature 1981;291:159-160.
26. Kurimoto Y, DeWeck AL, Dahinden CA. Interleukin-3 dependent mediator release in basophils triggered by C5a. J Exp Med 1989;170:467-79.
27. Leary AG, Ogawa M. Identification of pure and mixed basophil colonies in culture of human peripheral blood and bone marrow. Blood 1984;64:78-83.
28. Longley BJ, Morganroth GS, Teryll L, Ding TG, Anderson DM, Williams DE, Halaban R. Altered metabolism of mast cell growth factor (c-kit ligand) in cutaneous mastocytosis. N Engl J Med 1993;328:1302-07.
29. Lopez AF, Eglinton JM, Gillis D, Park LS, Clark S, Vadas MA. Reciprocal inhibition of binding between interleukin-3 and granulocyte-macrophage colony-stimulating factor to human eosinophils. Proc Natl Acad Sci (USA) 1989;86:7022-7.
30. Lopez AF, Eglington JM, Lyons AB, Tapley PM, To LB, Park LS, Clark SC, Vadas MA. Human interleukin-3 inhibits the binding of granulocyte macrophage colony stimulating factor and interleukin-5 to basophils and strongly enhances their functional properties. J Cell Physiol 1990;145:69-77.
31. Matsuda H, Coughlin MD, Bienenstock J, Denburg JA. Nerve growth

factor promotes human hemopoietic colony growth and differentiation. Proc natl Acad Sci (USA) 1988;85:6508-13.
32. Mayer P, Valent P, Schmidt G, Liehl E, Bettelheim P. The in-vivo effects of recombinant human interleukin-3: Demonstration of basophil differentiation factor, histamine producing activity and priming of GM-CSF responsive progenitors. Blood 1989;74:613-21.
33. Meininger CJ, Yano H, Rottapel R, Bernstein A, Zsebo KM, Zetter BR. The c-kit receptor ligand functions as a mast cell chemoattractant. Blood 1992;79:958-63.
34. Mitsui H, Furitsu T, Dvorak AM, Irani AMA, Schwartz LB, Inagaki N, Takei M, Ishizaka K, Zsebo KM, Gillis S, Ishizaka T. Development of human mast cells from umbilical cord blood cells by recombinant human and murine c-kit ligand. Proc Natl Acad Sci (USA) 1993;90:735-40.
35. Saito H, Hatake K, Dvorak AM, Leiferman KM, Donenberg AD, Arai N, Ishizaka K, Ishizaka T. Selective differentiation and proliferation of hemopoietic cells induced by recombinant human interleukins. Proc Natl Acad Sci (USA) 1988;85:2288-92.
36. Schleimer RP, Derse CP, Friedmann B, Gillis S, Plaut M, Lichtenstein LM, MacGlashan DW. Regulation of basophil mediator release by cytokines. J Immunol 1989;143:1310-7.
37. Schwartz LB, Irani AMA, Roller K, Castells MC, Schechter NM. Quantitation of histamine, tryptase, and chymase in dispersed human T- and TC mast cells. J Immunol 1987;138:2611-5.
38. Serafin WE, Austen KF. Mediators of immediate hypersensitivity reactions. New Engl J Med 1987;317:30-4.
39. Sillaber C, Bevec D, Ashman LK, et al. IL-4 regulates c-kit proto-oncogene product expression in human mast and myeloid progenitor cells. J Immunol 1991;147:4224-8.
40. Sillaber Ch, Geissler K, Kaltenbrunner R, Lechner K, Bettelheim P, Valent P. Transforming growth factor ß1 promotes IL-3 dependent differentiation of human basophils but inhibits IL-3 dependent differentiation of human eosinophils. Blood 1992;80:634-41.
41. Sperr WR, Czerwenka K, Mundigler G et al. Specific activation of human mast cells by the ligend for c-kit: comparisson between lung-, uterus and heart mast cells. Int Arch Allergy & Immunol 1993;102:170-5.
42. Thompson HL, Metcalfe DD, Kinet JP. Early expression of high affinity receptor for immunoglobulin E (FcERI) during differentiation of mouse mast cells and human basophils. J Clin Invest 1990;85:1227-33.
43. Tsai M, Takeishi T, Thompson H, Langley KE, Zsebo KM, Metcalfe DD, Geissler EN, Galli SJ. Induction of mast cell proliferation, maturation and

heparin synthesis by the rat c-kit ligfand, stem cell factor. Proc Natl Acad Sci (USA) 1991;88:6382-6.
44. Tsuda T, Wong D, Dolovich J, Bienenstock J, Marshall J, Denburg JA. Synergistic effect of nerve growth factor and granulocyte-macrophage colony stimulating factor on human basophilic cell differentiation. Blood 1991;77:971-9.
45. Valent P. The riddle of the mast cells: kit ligand as the missing link ? Immunol Today 1994;15:111-4.
46. Valent P, Besemer J, Muhm M, Majdic O, Lechner K, Bettelheim P. Interleukin 3 activates human blood basophils via high-affinity binding sites. Proc Natl Acad Sci (USA) 1989;86:5542-6.
47. Valent P, Besemer J, Sillaber C et al. Failure to detect interleukin-3 binding sites on human mast cells. J Immunol 1990;145:3432-7.
48. Valent P, Bettelheim P. Cell surface structures on human basophils and mast cells: Biochemical and functional characterization. Adv Immunol 1992;52:333-423.
49. Valent P, Schmidt G, Besemer J, Mayer P, Liehl E, Hinterberger W, Lechner K, Maurer D, Bettelheim P. Interleukin-3 is a differentiation factor for human basophils. Blood 1989;73:1763-9.
50. Valent P, Spanblöchl E, Sperr WR, et al. Induction of differentiation of human mast cells from bone marrow and peripheral blood mononuclear cells by recombinant human stem cell factor (SCF) / kit ligand (KL) in long term culture. Blood 1992;80:2237-45.
51. Zsebo KM, Williams DA, Geissler EN, et al. Stem cell factor is encoded at the Sl locus of the mouse and is the ligand for the c-kit tyrosine kinase receptor. Cell 1990;63:213-228.

11

Keratinocyte-derived Modulators of Mast Cell Growth

Shoso Yamamoto, Koji Nakamura, Toshihiko Tanaka, Eishin Morita and Yoshikazu Kameyoshi

Department of Dermatology, Hiroshima University School of Medicine, Hiroshima

Mast cells have been recognized as cells playing an important role in the pathophysiology not only of allergic and immunological diseases but also of various other diseases. They participate in physiological and pathological processes through the production and release of biologically active substances such as preformed and newly generated mediators or cytokines. Therefore, an increase in the number of mast cells at the sites of lesions may affect the pathophysiology of the diseases. An increase in number of mast cells has been observed in various skin diseases (table 1) (1). However, little is known about the mechanism of increase in these diseases. It is not easy to study the mechanism of mast cell growth in the diseases, because there is lacking a suitable experimental model for investigating mast cell proliferation in the pathological conditions.

An increase in the number of mast cells has been observed at the sites of cancer. Mikhail et al. reported high mast cell counts in basal cell carcinoma and in the stroma of squamous cell carcinoma in skin (2). Claudatus et al. have likewise observed that mast cells are consistently found in both types of tumor, tending to increase as the degree of malignancy worsens (3). These facts suggest that tumor

Table 1. Skin diseases with mast cell hyperplasia

Atopic dermatitis
Allergic contact dermatitis
Lichen simplex chronicus
Lichen planus
Pemphigus vulgaris
Granulation tissue of healing wounds
Neurofibroma
Urticaria pigmentosa
Carcinoma

cells may play an important role in the accumulation of the mast cells at the site of the tumor. Therefore, we chose keratinocyte-derived squamous cell carcinoma in mouse as an experimental model for investigating the mechanism of mast cell growth in pathological conditions.

It is known that induction of mast cell growth in mice is mediated by various cytokines or other factors, such as interleukin 3 (IL-3) (4), IL-4 (5), IL-9 (6), IL-10 (7), stem cell factor (SCF) that is a ligand for c-*kit* (8, 9) and nerve growth factor (NGF) (10). In mice, these factors, separately or together, may regulate mast cell growth. Based on these facts, we investigated whether or not these factors are responsible for the increase of mast cell number in the cancer tissues. The results of the present study indicate that tumor cells may produce and secrete some factor which enhances the fibroblast-dependent mast cell growth and this factor may be different from already known factors for mast cell growth.

INCREASE OF MAST CELL NUMBER IN SQUAMOUS CELL CARCINOMA OF MURINE SKIN

In order to investigate the mechanism of mast cell growth in tumor, a tumor cell line was established from squamous cell carcinoma in murine skin. Tumors were induced chemically on the dorsal skin of CBA/J male mice by applying 3-methylcholanthrene and 12-o-tetradecanoylphorbol-13-acetate.

From one of the ensuing skin tumors, a squamous cell carcinoma cell line was established and designated KCMH-1. All of the cultured KCMH-1 cells were positively stained with anti-keratin antibody. On electron-microscopic examination, numerous cytoplasmic tonofilaments and characteristic desmosomes were observed in cultured KCMH-1 cells, suggesting that the KCMH-1 cells were derived from keratinocytes (11).

At the sites of the inoculation of 5×10^5 KCMH-1 cells into the dorsal skin of normal healthy CBA/J male mice, the tumors with approximately 9 mm in diameter developed 40 days after the inoculation. As shown in table 2, a significant increase in the number of mast cells was observed at the sites of tumors (11). These results suggest that KCMH-1 cells may secrete some factor which promotes mast cell growth in vivo.

Table 2. Increase of mast cell population in tumors developed at skin sites of KCMH-1 cell inoculation

Mast cells/mm^2	
Tumor	Control
141.4 ± 54.8	52.3 ± 11.0
(n=4)	(n=5)

KCMH-1 cells (5×10^5 cells/site) were inoculated into the dorsal skin of healthy CBA/J mice, and biopsy was carried out on the 40th day after inoculation.

ENHANCEMENT OF FIBROBLAST-DEPENDENT MAST CELL GROWTH IN VITRO BY CONDITIONED MEDIUM OF MURINE SQUAMOUS CELL CARCINOMA CELLS

It is well known that keratinocytes produce and secrete various cytokines including IL-3 (12) which is one of the most important factors for mast cell growth in mice. In order to investigate whether KCMH-1 cells secrete mast cell growth factors, bone marrow-derived mast cells (BMMC) obtained from CBA/J male mice were cultured in the presence of the conditioned medium of KCMH-1 cells (TCM).

The BMMC disappeared within 2 days even in the presence of 50% TCM, suggesting that an insufficient amount of mast cell growth factor directly acting on BMMC was contained in the TCM.

However, when BMMC were cultured on NIH/3T3 fibroblast monolayer (13), the alcian blue positive cells (mast cells) increased in number in the presence of the TCM in a dose dependent manner. An increase in total histamine content in the culture dishes was also observed in the presence of the TCM in a dose dependent manner (table 3) (11). In the mast cell growth induced by TCM on the fibroblast monolayer, the adherence of BMMC to fibroblasts may be essential, because BMMC failed to survive even in the presence of the conditioned medium of NIH/3T3 fibroblasts cultured with 50% TCM for 2 days.

MAST CELL GROWTH FACTORS IN CONDITIONED MEDIUM OF MURINE SQUAMOUS CELL CARCINOMA CELLS

The results described above suggest that the TCM supports mast cell growth synergistically with NIH/3T3 fibroblasts. In mice, IL-3 (4), IL-4 (5), IL-9 (6), IL-10 (7), SCF (8,9,14) and NGF (10) promote proliferation and differentiation of mast cells separately or together. Therefore, we investigated whether the TCM contains these factors. As shown in table 4, addition of rmIL-3, rmIL-4 or rmIL-10 to the BMMC-NIH/3T3 fibroblast coculture caused synergistic enhancement

Table 3. Enhancement of fibroblast-dependent mast cell growth by TCM in vitro

Concentration of TCM (%)	Number of mast cells ($\times 10^4$/dish)	Histamine content (ng/dish)
0	7.5 ± 2.1	54.7 ± 16.9
6.25	27.2 ± 7.6	84.6 ± 13.6
12.5	38.1 ± 28.3	92.7 ± 14.8
25	39.1 ± 24.6	169.7 ± 22.1
50	36.2 ± 14.0	159.8 ± 3.7

BMMC derived from CBA/J mice (5×10^5 cells/dish) were cultured on NIH/3T3 fibroblast monolayer for 14 days in the presence of TCM at various concentrations. Data represent mean ± SD of three separate experiments. Each experiment was carried out in triplicate.

of mast cell growth, and the activity of each factor for mast cell growth was neutralized by monoclonal antibody specific for the corresponding factor respectively. However, these antibodies failed to abolish mast cell growth induced by the TCM in the coculture, suggesting an insufficient amount of IL-3, IL-4 and IL-10 for mast cell growth in the TCM (11). NGF showed no effect on the mast cell growth in the coculture under these experimental conditions. In these experiments, histamine was quantified as a marker of mast cell proliferation, because the number of alcian blue positive cells (mast cells) paralleled the content of histamine in the coculture.

The mast cell growth was also enhanced by the addition of rmSCF or rmIL-9 in the coculture. However, in the present study, it was impossible to analyze the activity of soluble SCF in the culture medium by antibody specific for SCF without affection on membrane-associated SCF of fibroblasts. No commercially available mouse monoclonal antibody specific for IL-9 could be obtained. Therefore, in the assay of mast cell proliferation induced by SCF or IL-9,

Table 4. Effect of anti-IL-3, anti-IL-4 and anti-IL-10 antibodies on TCM-induced mast cell growth in coculture

Factor	Antibody (μg/ml)	Histamine content (ng/well)
IL-3 experiment	Anti-mIL-3 mAb	
Control medium	0	18.44 ± 1.33
	1	16.80 ± 2.02
TCM (50%)	0	57.70 ± 6.52
	1	49.77 ± 2.36
rmIL-3 (10U/ml)	0	53.48 ± 6.63
	1	19.00 ± 0.80
IL-4 experiment	Anti-mIL-4 mAb	
Control medium	0	11.98 ± 1.24
	10	14.59 ± 1.17
TCM (50%)	0	29.43 ± 3.20
	10	29.00 ± 2.46
rmIL-4 (100ng/ml)	0	21.36 ± 2.21
	10	15.76 ± 1.08
IL-10 experiment	Anti-mIL-10 mAb	
Control medium	0	14.57 ± 1.29
	1	15.56 ± 2.28
TCM (50%)	0	27.96 ± 1.41
	1	32.77 ± 2.18
rmIL-10 (5ng/ml)	0	53.46 ± 4.14
	1	21.72 ± 2.26

The medium containing TCM or each factor was incubated with antibody specific for each factor for 1 hr at 37°C before culture. BMMC (1×10^5 cells/well) were cultured on NIH/3T3 fibroblast monolayer in the medium for 10 days. Data shown in the table are mean ± SD of triplicate cultures.

[³H]thymidine uptake was quantified as a marker of cell proliferation in BMMC suspension culture. BMMC responded to rmSCF in [³H]thymidine uptake and the combination of rmSCF with rmIL-9 markedly enhanced this response. However, the TCM alone or in combination with rmSCF failed to increase [³H]thymidine uptake in BMMC, suggesting that the TCM contains an insufficient amount of SCF and IL-9 for mast cell growth (table 5) (11). These results strongly suggest the possibility that the mast cell growth induced by the TCM in mast cell-fibroblast coculture is mediated by other factors in the TCM different from the already known mast cell growth factors such as IL-3, IL-4, IL-9, IL-10, SCF or NGF.

Another possibility in the accumulation of mast cells at the sites of tumors is mast cell migration. However, it seems unlikely that the migration of the mast cells is a central mechanism of the cell accumulation, because mature mast cells are unresponsive to a wide range of chemotactic factors (15).

INDUCTION OF PROLIFERATION OF MAST CELLS DERIVED FROM WBB6F$_1$-W/W^v MICE BY CONDITIONED MEDIUM OF MURINE SQUAMOUS CELL CARCINOMA CELLS

The results outlined above indicate that in CBA/J mice mast cell growth induced by the TCM needed the adherence of BMMC to NIH/3T3 fibroblasts. This raises the possibility that the TCM may induce the increase of SCF expression and/or the expression of some factor different from SCF for mast cell growth on the fibroblast membrane.

In the preliminary experiments, although SCF mRNA and SCF protein were detected in NIH/3T3 fibroblasts by Northern blotting analysis and Western blotting analysis respectively, no increase of mRNA for SCF and of SCF protein was observed in the fibroblasts treated with TCM for 48 hr. Therefore, the possibility was investigated whether another factor different from SCF for mast cell growth was induced by the TCM on the membrane of the fibroblasts using BMMC obtained from WBB6F$_1$-+/+ mice which express receptors for

Table 5. [³H]thymidine uptake in BMMC of CBA/J mice induced by TCM, rmSCF, rmIL-9 or their combinations.

Factors	[³H]thymidine uptake (cpm)
Control medium	80.7 ± 27.2
TCM (50%)	73.3 ± 18.0
rmSCF (10ng/ml)	3267.3 ± 324.9
rmIL-9 (1.0ng/ml)	78.0 ± 8.7
rmSCF (10ng/ml) + rmIL-9 (1.0ng/ml)	20143.0 ± 992.7
rmSCF (10ng/ml) + TCM (50%)	3584.3 ± 30.6

1×10^4 BMMC were cultured in the medium containing the factors for 48hr. The cultured cells were pulsed with 0.5 µCi [³H]thymidine for the last 12hr. Data represent the mean ± SD of triplicate cultures.

Table 6. Enhancement of fibroblast-dependent mast cell growth induced by TCM in WBB6F$_1$-+/+ and WBB6F$_1$-W/W^v mice

Mice	TCM (%)	Culture periods (days)	Number of mast cells (x10^4/well)
WBB6F$_1$-+/+			
	30	1	1.88 ± 0.72
	30	6	2.56 ± 0.62
	30	14	5.43 ± 0.18
	0	1	1.36 ± 0.28
	0	6	0.63 ± 0.27
	0	14	0.17 ± 0.06
WBB6F$_1$-W/W^v			
	30	1	1.88 ± 0.52
	30	6	1.68 ± 0.29
	30	14	1.72 ± 0.25
	0	1	1.87 ± 0.52
	0	6	0.18 ± 0.12
	0	14	0.03 ± 0.01

BMMC derived from WBB6F$_1$-+/+ mice or WBB6F1-W/W^v mice (2x10^4 cells/well) were cultured on NIH/3T3 fibroblast monolayer for various duration in the presence or absence of 30% TCM. Data represent mean ± SD in tetracate.

SCF and WBB6F$_1$-W/W^v mice which lack functional receptors for SCF. When BMMC derived from WBB6F$_1$-+/+ mice were cultured on NIH/3T3 fibroblast monolayer, they gradually decreased in number throughout the culture period, but still survived on day 14 of the coculture. In the absence of fibroblasts, they disappeared completely by the 6th day of the culture, suggesting that the +/+ mast cells needed the fibroblasts for their survival. By contrast, the addition of TCM to the +/+ mast cell-NIH/3T3 fibroblast coculture caused a marked increase in the number of mast cells (table 6), as seen in the case of CBA/J mast cells.

Meanwhile, the addition of the TCM to the W/W^v mast cell-NIH/3T3 fibroblast coculture system caused survival of the mast cells throughout the culture for 14 days, although W/W^v mast cells quickly decreased in number in the absence of the TCM. The number of adherent mast cells on the fibroblast monolayer was relatively constant throughout the coculture for 14 days in the presence of the TCM, in spite of the removal of nonadherent mast cells at each medium change every 2 days.

This fact suggest that W/W^v mast cells may proliferate on NIH/3T3 fibroblast monolayer in the presence of the TCM, suggesting that the TCM may induce some factor different from SCF for mast cell growth on the fibroblast membrane.

CONCLUSION

In this study, in order to know the mechanism of mast cell proliferation and differentiation, we investigated mast cell increase at the sites of inoculation of keratinocyte-derived squamous cell carcinoma cell line (KCMH-1) cells in CBA/J mice. Significant increase in the number of mast cells was observed at the sites of tumors developed at the sites of inoculation of KCMH-1 cells. Through in vitro experiments, enhancement of mast cell growth was observed by culturing bone marrow-derived mast cells (BMMC) derived from CBA/J mice on NIH 3T3 fibroblast monolayer in the presence of conditioned medium obtained from KCMH-1 (TCM). The activities of the known factors for mast cell growth, such as IL-3, IL-4, IL-9, IL-10, SCF or NGF were not detected in the TCM. The mast cell growth induced by the TCM in vitro required the adherence of mast cells to NIH/3T3 fibroblasts. The treatment of NIH/3T3 fibroblasts with the TCM did not cause an increase of SCF mRNA and SCF protein of NIH/3T3 fibroblasts. Moreover, the TCM supported the survival of mast cells derived from W/W^v mice, in which c-*kit* receptor is functionally deficient, on the NIH/3T3 fibroblast monolayer. These results suggest that transformed keratinocytes may produce some factor which induces expression of unknown factor, different from SCF, for mast cell growth on fibroblast membrane. The mast cell growth observed in this study may be due to a novel mechanism of mast cell proliferation and differentiation.

REFERENCES

1. Lever WF, Schaumburg-Lever G. Chapter 5, Morphology of the cells in the dermal infiltrate. In: *Histopathology of the Skin* , (17th Edition). Philadelphia: J.B. Lippincott Company; 1990: 55-64.
2. Mikhail GR, Miller-Milinska A. Mast cell population in human skin. *J Invest Dermatol* 1964; 43: 249-254.
3. Claudatus JC Jr, d'Ovidio R, Lospalluti M, Meneghini CL. Skin tumors and reactive cellular infiltrate: Further studies. *Acta Derm Veneol (Stockh)* 1986; 66: 29-34.
4. Ihle JN, Keller J, Oroszlan S, et al. Biological properties of homogeneous interleukin 3. I. Demonstration of WEHI-3 growth factor activity, mast cell growth factor activity, P cell-stimulating factor activity, colony-stimulating factor activity, and histamine-producing cell-stimulating factor activity. *J Immunol* 1983; 131: 282-287.
5. Hamaguchi Y, Kanakura Y, Fujita J, et al. Interleukin 4 as an essential factor for in vitro clonal growth of murine connective tissue type mast cells. *J Exp Med* 1987; 165: 268-273.
6. Hultner L, Druez C, Moeller J, et al. Mast cell growth-enhancing activity (MEA) is structurally related and functionally identical to the novel mouse T cell growth factor P40/TCGF III (interleukin 9). *Eur J Immunol* 1990; 20: 1413-1416.
7. Thompson-Snipes L, Dhar V, Bond MW, et al. Interleukin 10: A novel stimulatory factor for mast cells and their progenitors. *J Exp Med* 1991; 173: 507-510.

8. Zsebo KM, Williams DA, Geissler EN, et al. Stem cell factor (SCF) is encoded at the *Sl* locus of the mouse and is the ligand for the c-*kit* tyrosine kinase receptor. *Cell* 1990; 63: 213-224.
9. Zsebo KM, Wypych J, McNiece IK, et al. Identification, purification, and biological characterization of hematopoietic stem cell factor from buffalo rat liver-conditioned medium. *Cell* 1990; 63:195-201.
10. Matsuda H, Kannan Y, Ushio H, et al. Nerve growth factor induces development of connective tissue-type mast cells in vitro from murine bone marrow cells. *J Exp Med* 1991; 174: 7-14.
11. Nakamura K, Tanaka T, Morita E, Yamamoto S. Enhancement of fibroblast-dependent mast cell growth by conditioned medium of keratinocyte-derived squamous cell carcinoma cells in mouse. *Arch Dermatol Res* (in press).
12. Luger TA. Epidermal cytokines. *Acta Derm Veneol (Stockh)* 1989; 69:61-76.
13. Levi-Schaffer F, Austen KF, Gravallese PM, Stevens RL. Coculture of interleukin 3-dependent mouse mast cells with fibroblasts results in a phenotypic change of the mast cells. *Proc Natl Acad Sci (U.S.A.)* 1986; 83: 6485-6488.
14. Tsuji K, Zsebo KM, Ogawa M. Murine mast cell colony formation supported by IL-3, IL-4, and recombinant rat stem cell factor, ligand for c-*kit*. *J Cell Physiol* 1991; 148: 362-369.
15. Czarnetzki BM, Wullenweber I. In vitro migratory response of rat peritoneal macrophages and mast cells towards chemotactic factors and growth factors. *J Invest Dermatol* 1988; 91: 224-227.

Biological and Molecular Aspects of Mast Cell and Basophil Differentiation and Function,
edited by Y. Kitamura, S. Yamamoto, S.J. Galli, and M.W. Greaves. Raven Press, Ltd., New York © 1995.

12

ADHESION MOLECULES AND THEIR RELEVANCE IN UNDERSTANDING THE BIOLOGY OF THE MAST CELL

Peter J. Bianchine, M.D., and Dean D. Metcalfe, M.D.

Allergic Diseases Section, Laboratory of Clinical Investigation, National Institute of Allergy and Infectious Diseases, National Institutes of Health, Bethesda, MD 20892

The mature differentiated mast cell exists exclusively within tissues under non-pathological conditions. This characteristic clearly distinguishes it from other hematopoietically derived cells, such as basophils, neutrophils, and eosinophils. Mast cells are found throughout the connective tissues of almost every organ system in the body. They congregate around nerves, blood vessels, and lymphatic vessels in a remarkably consistent fashion. To these observations must also be added the fact that mast cell phenotype varies morphologically, histochemically, and ultrastructurally with its distribution. If it is assumed that mast cell distribution and heterogeneity relate to the biologic function of the cell, then to clearly understand the basis of mast cell biology it is necessary to understand the intimate interactions between mast cells and the extracellular connective tissue matrix components.

MAST CELL ADHESION TO LAMININ

The distribution of mast cells along basement membranes directed us to first focus our attention on the possibility that mast cells might bind to laminin. We first observed that the mouse PT18 mast cell line spontaneously adhered to surfaces coated with laminin, while bone marrow derived cultured mast cells (BMCMC) which were IL-3 dependent had to be activated with PMA to adhere (1). Mast cell attachment was observed to be accompanied by cell spreading and a redistribution of cytoplasmic granules. Adherence to laminin was inhibited by antibodies to laminin or antibodies to laminin receptors. Subsequently, laminin fragments were also observed to be a chemotactic stimulus for mast cells (2). A synthetic peptide containing the sequence IKVAV, derived from the A chain of the laminin molecule was particularly active in mast cell attachment and migration (3).

These observations next led us to explore the physiologic relevance of mast cell adhesion. First, we demonstrated that aggregation of $Fc_\varepsilon RI$ on BMCMC was a highly sensitive physiologic stimulus for this attachment (4). This $Fc_\varepsilon RI$-dependent adherence occurred at threefold log concentrations of antigen less than that required for histamine release. At higher concentrations of antigen associated with degranulation, the time course for adhesion was shown to be more prolonged than that required for histamine release. Culture of mast cells with TGF-β enhanced IgE-mediated adhesion of mast cells to laminin.

Adhesion was also shown to be dependent upon calcium and to be temperature-dependent (3). Dibutyryl cAMP inhibited both adherence and histamine release. Staurosporin which inhibits protein kinase C inhibited adherence induced by calcium ionophore. These studies on laminin clearly demonstrated that mast cells had the capability to adhere to matrix components under physiologic conditions in a regulated process separate from degranulation.

ADHESION OF MAST CELL PROGENITORS

It is now accepted that mast cells may be cultured from bone marrow-derived pluripotential stem cells. Thus, human CD34+ cells give rise to mast cells in the presence of mast cell growth factors including IL-3 and c-kit ligand, or stem cell factor (SCF) (5,6). SCF in particular directs mast cell maturation in both murine (7) and human (6) systems. Mast cell precursors express $Fc_\varepsilon RI$ and $Fc_\gamma RII/III$ prior to granulation (8,9).

These observations led us to question whether mast cell precursors would also adhere to a tissue matrix component, specifically laminin. In this way, SCF-dependent mast cell precursors as identified in peripheral blood (10) could localize in specific tissues. We thus isolated mast cell precursors from mouse bone marrow cultures after 1 week of culture with IL-3 by cell sorting for $Fc_\varepsilon RI^+$ cells. These precursors also exhibited a dose-dependent adherence to laminin following activation (4).

ADHERENCE TO FIBRONECTIN AND VITRONECTIN

Following demonstration that mast cells could interact with laminin, it was reasonable to hypothesize that mast cells would also adhere to other matrix components such as fibronectin and vitronectin. As expected, activated mast cells were shown to adhere to both fibronectin (11) and vitronectin (12). As with laminin, adherence to fibronectin required cell activation with PMA or occurred following aggregation of $Fc_\varepsilon RI$. In both interactions, adhesion was blocked with

RGD-containing peptides, implicating adhesion mediated by members of the integrin family. Adhesion to vitronectin was inhibited specifically by antibody to α_v, and adhesion to fibronectin with antibody to α_3.

STIMULUS DEPENDENT VS. SPONTANEOUS ADHESION

Early studies with various mast cell lines revealed that some of these lines exhibited high rates of spontaneous adhesion to laminin and fibronectin. For instance, without activation, PT18 mast cells readily adhered to laminin (1). In contrast, IL-3 dependent BMCMC would only adhere to laminin or fibronectin following activation with PMA or following Fc_eRI aggregation. However, when the interaction between BMCMC and vitronectin was examined, it was discovered that BMCMC spontaneously adhered to vitronectin (12). Adhesion was not increased further when PMA was added or IgE receptors were aggregated. Spontaneous adhesion persisted over 180 minutes and was not associated with histamine release.

SCF-INDUCED MAST CELL ADHESION

Neither PMA-induced mast cell adhesion, nor mast cell adhesion following aggregation of Fc_eRI is a candidate for the stimulus that directs mast cell adherence to matrix under normal physiologic conditions. In fact, a number of cytokines did not promote mast cell adhesion to laminin including platelet-derived growth factor, IFN-γ, TNF, GM-CSF, epidermal growth factor, IL-1, IL-2, IL-3, IL-4, endothelial cell growth factor -α, and endothelial cell growth factor -β (4).

We were thus interested in identifying a factor that would regulate mast cell adhesion under physiologic conditions and in the absence of antigen. SCF, because of its properties and sites of synthesis in connective tissues seemed to be a strong candidate for such a factor. Indeed, SCF did turn out to be a potent activator of mast cell adhesion.

SCF was found to promote the adhesion of IL-3-dependent BMCMC to fibronectin in a dose response fashion, with 50-60% of BMCMC adhering to fibronectin at a concentration of 10 ng/ml of SCF (13). In fact, near maximal adhesion was provoked by 0.5 to 2.5 ng/ml which is the approximate range of SCF concentrations in plasma. SCF was shown to mediate adhesion through interaction with c-kit. RGD-containing peptides blocked adhesion, thereby implicating integrins (Figure 1). Genistein, which inhibits tyrosine kinase activity, partially inhibited SCF-induced adhesion.

Mast cell distribution in tissues appears to depend on the homing of mast cell progenitors to specific locations. Receptor-mediated recognition of extracellular components is one critical mechanism in this process. SCF, which is a major physiologic stimulus for mast cells, and is produced by resident connective tissue

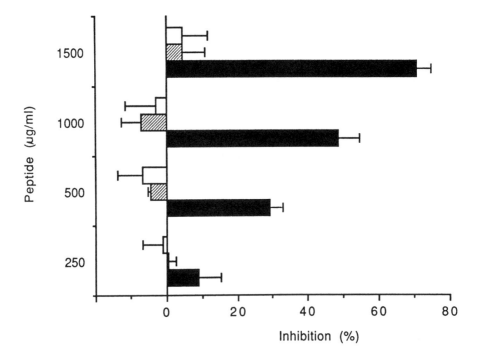

FIG. 1. Inhibition of SCF-induced mast cell adhesion to FN by peptides. BMCMC from BALB/c mice were incubated with 10 ng/ml of SCF in wells coated with 10 µg/ml FN. Increasing concentrations of GRGDSP (solid bars), GRGESP (open bars), or EILDVSPST (shaded bars) were added before SCF. Adhesion in the wells without peptide was taken as 100% and the percentage of inhibition was calculated. Each point represents mean ± SEM from three independent experiments, each performed in duplicate. Statistical analysis performed on nontransformed data (percentage of adhesion) showed significant differences at p = 0.05 between the control and 500, 1000, and 1500 mg/ml of GRGDSP. Reprinted from reference 13 with permission.

cells, appears to be a second feature of the microenvironment governing mast cell distribution. It follows that up-regulation or down-regulation of SCF in pathologic conditions could provide one means by which mast cell distribution can be governed.

INFLUENCES OF ADHESION ON MAST CELL RESPONSES

Interactions between cells and matrix components have been reported to alter the biology of the involved cell. For instance, embryonic chick retinal neurons adhering to vitronectin exhibit neurite outgrowth (15). As a first attempt to explore the possibility that matrix influences mast cells, we chose to examine whether mast cells after attachment to matrix components would exhibit migration as assessed with time-lapse videography. This possibility was suggested to us by the observation that mast cells found along basement membranes appear spindle shaped, reminiscent of the appearance of macrophages exhibiting locomotion. As expected, $Fc_\varepsilon RI$ activated mast cells adhered to laminin, fibronectin, and matrigel, continually exhibiting periods of movement as round cells with small pseudopodia interspaced with periods of flattening (Figure 2). The mean velocity of BMCMC on laminin, fibronectin, and matrigel was similar and averaged approximately 180 µm/hr.

While the demonstration of mast cell movement on matrix components following aggregation of IgE receptors suggested an effect of matrix itself on the mast cell, it could not be conclusively shown that this effect was not entirely a consequence of the aggregation of IgE receptors and that the matrix was simply passive. To prove that the interactions between mast cell receptors for matrix and matrix components could alter mast cell behavior, we examined the consequences of spontaneous adhesion to vitronectin where no activating signal is required on the proliferative rate of mast cells. As reported (12), there was a 41% increase in the rate of mast cell proliferation over a 96 hour period following adhesion to vitronectin in the presence of IL-3. This effect was prevented by the addition of RGD-containing peptides. More recently we have reported that the spontaneous adhesion of mast cells to vitronectin is followed by the rapid phosphorylation of multiple intracellular proteins (16).

It is now increasingly clear that interactions between mast cells and matrix components alter mast cell behavior. In addition to the effect on proliferation, mast cell adhesion has been reported to enhance $Fc_\varepsilon RI$-dependent mast cell histamine release (17), and synergistically regulate tyrosine phosphorylation of proteins including the $Fc_\varepsilon RI$-induced tyrosine phosphorylation of focal adhesion kinase (18,19).

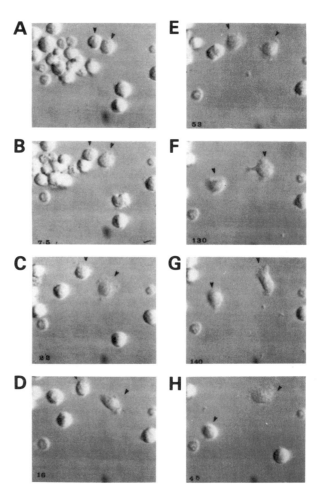

FIG. 2. (a-h) Flattening-ruffling behavior with migration exhibited by two BMCMC on laminin substratum following $Fc_\varepsilon RI$-mediated activation. Pictures were selected to illustrate behavior on the substratum. Times shown in the lower left hand corner of each panel indicate minutes between successive photographs. Arrows indicate cells demonstrating flattening-ruffling motion. Reprinted from reference 14 with permission.

SUMMARY

The interactions between mast cells and connective tissue matrix components clearly have profound influences on the targeting of mast cell progenitors to specific locations, the distribution of mast cell subsets, and the biologic responsiveness of mast cells in tissues. In general these interactions are mediated by integrins. Adhesion of mast cells to matrix may be both stimulus-dependent and stimulus independent. Our preliminary evidence suggests that stimulus-dependent adhesion of mast cells via integrins takes place through a change in the avidity of the integrin receptor for its matrix ligand via "inside-out signalling" (20). The engagement of matrix components by mast cells clearly up-regulates mast cell processes including intracellular protein phosphorylation, thymidine incorporation, histamine release, and cell motility. It is reasonable to expect these matrix - mast cell interactions will also result in an alteration in morphology (1,6,21) cytokine synthesis, the synthesis of granule matrix components, cell survival and apoptosis as has been described for mast cells following IL-3 withdrawal (22). Indeed, these interactions between mast cells and their surrounding matrix environment may eventually be shown to be the major regulatory stimuli for fine-tuning the biologic responsiveness of the mature differentiated mast cell.

ACKNOWLEDGEMENTS

The authors thank Mrs. Belinda Richardson for her editorial support in the preparation of this manuscript.

REFERENCES

1. Thompson HL, Burbelo PD, Segui-Real B, Yamada Y, Metcalfe DD. Laminin promotes mast cell attachment. *J Immunol* 1989;143:2323-7.
2. Thompson HL, Burbelo PD, Yamada Y, Kleinman HK, Metcalfe DD. Mast cells chemotax to laminin with enhancement after IgE-mediated activation. *J Immunol* 1989;143:4188-92.
3. Thompson HL, Burbelo PD, Yamada Y, Kleinman HK, Metcalfe DD. Identification of an amino acid sequence in the laminin A chain mediating mast cell attachment and spreading. *Immunology* 1991;72:144-9.
4. Thompson HL, Burbelo PD, Metcalfe DD. Regulation of adhesion of mouse bone marrow-derived mast cells to laminin. *J Immunol* 1990;145:3425-31.

5. Kirshenbaum AS, Kessler SW, Goff JP, Metcalfe DD. Demonstration of the origin of human mast cells from CD34+ bone marrow progenitor cells. *J Immunol* 1991;146:1410-5.
6. Kirshenbaum AS, Goff JP, Kessler SW, Mican JM, Zsebo KM, Metcalfe DD. Effect of IL-3 and stem cell factor on the appearance of human basophils and mast cells from CD34+ pluripotent progenitor cells. *J Immunol* 1992;148:772-7.
7. Tsai M, Takeishi T, Thompson HL, et al. Induction of mast cell proliferation, maturation, and heparin synthesis by the rat c-kit ligand, stem cell factor. *Proc Natl Acad Sci USA* 1991;88:6382-6.
8. Thompson HL, Metcalfe DD, Kinet J-P. Early expression of high-affinity receptor for immunoglobulin E ($Fc_\varepsilon RI$) during differentiation of mouse mast cells and human basophils. *J Clin Invest* 1990;85:1227-33.
9. Rottem M, Barbieri S, Kinet J-P, Metcalfe DD. Kinetics of the appearance of $Fc_\varepsilon RI$-bearing cells in interleukin-3-dependent mouse bone marrow cultures: correlation with histamine content and mast cell maturation. *Blood* 1992;79:972-80.
10. Rottem M, Hull G, Metcalfe DD. Demonstration of differential effects of cytokines on mast cells derived from murine bone marrow and peripheral blood mononuclear cells. *Exp Hematol*, in press.
11. Dastych J, Costa JJ, Thompson HL, Metcalfe DD. Mast cell adhesion to fibronectin. *Immunology* 1991;73:478-84.
12. Bianchine PJ, Burd PR, Metcalfe DD. IL-3-dependent mast cells attach to plate-bound vitronectin: Demonstration of augmented proliferation in response to signals transduced via cell surface vitronectin receptors. *J Immunol* 1992;149:3665-71.
13. Dastych J, Metcalfe DD. Stem cell factor induces mast cell adhesion to fibronectin. *J Immunol* 1994;152:213-9.
14. Thompson HL, Thomas L, Metcalfe DD. Murine mast cells attach to and migrate on laminin-, fibronectin-, and matrigel-coated surfaces in response to $Fc_\varepsilon RI$-mediated signals. *Clin Exp Allergy* 1992;23:270-5.
15. Neugebauer KM, Emmett CJ, Venstrom KA, Reichardt LF. Vitronectin and thrombospondin promote retinal neurite outgrowth: developmental regulation and role of integrins. *Neuron* 1991;6:345-51.
16. Bianchine PJ, Paolini R, Kinet J-P, Metcalfe DD. Mast cell adhesion to vitronectin is sufficient to phosphorylate focal adhesion kinase. *J Allergy Clin Immunol* 1994;93:226 (380A).
17. Hamawy MM, Oliver C, Mergenhagen SE, Siraganian RP. Adherence of rate basophilic leukemia (RBL-2H3) cells to fibronectin-coated surfaces enhances secretion. *J Immunol* 1992;149:615-21.
18. Hamawy MM, Mergenhagen SE, Siraganian RP. Cell adherence to fibronectin and the aggregation of the high affinity immunoglobulin E receptor synergistically regulate tyrosine phosphorylation of 105-115-kDa proteins. *J Biol Chem* 1993;268:5227-33.

19. Hamawy MM, Mergenhagen SE, Siraganian RP. Tyrosine phosphorylation of pp125^{FAK} by the aggregation of high affinity immunoglobulin E receptors requires cell adherence. *J Biol Chem* 1993;268:6851-4.
20. Dastych J, Wyczolkowska J, Metcalfe DD. IgE-crosslinking alters the avidity of an alpha-5 containing integrin receptor on murine mast cells for fibronectin. *J Allergy Clin Immunol* 1994;93:226 (379A).
21. Rottem M, Goff JP, Albert JP, Metcalfe DD. The effects of stem cell factor on the ultrastructure of Fc$_e$RI$^+$ cells developing in IL-3-dependent murine bone marrow-derived cell cultures. *J Immunol* 1993;151:4950-63.
22. Mekori YA, Oh CK, Metcalfe DD. IL-3-dependent murine mast cells undergo apoptosis on removal of IL-3: Prevention of apoptosis by c-kit ligand. *J Immunol* 1993;151:3775-84.

Biological and Molecular Aspects of Mast Cell and Basophil Differentiation and Function,
edited by Y. Kitamura, S. Yamamoto, S.J. Galli, and
M.W. Greaves. Raven Press, Ltd., New York © 1995.

13

Mouse Mast Cell Proteases

John E. Hunt and Richard L. Stevens

Department of Medicine, Harvard Medical School, and Department of Rheumatology and Immunology, Brigham and Women's Hospital, Boston, Massachusetts 02115

Mast cells and their products are crucial links in the chain of events that lead to allergy and inflammation. While attention has focused on the proinflammatory mediators released from activated mast cells (e.g. histamine, leukotrienes, prostaglandin D_2, and cytokines), there has been an increasing awareness of the potential roles of the mast cell's granule serine proteases and exopeptidases in IgE-mediated immune responses. It is now apparent that mast cells exist *in vivo* and *in vitro* as a heterogeneous family of effector cells. Despite the development in the early 1980s of culture techniques to obtain large numbers of pure populations of non-transformed and transformed mouse mast cells (38,39,42-44, 46,49,55,62) and the discovery of mast cell-deficient mice (27,28), the lack of mast cell-specific markers hindered an understanding of the factors that regulate their growth, differentiation, and maturation. Mast cell proteases are enzymatically active at neutral pH and often represent >50% of the total protein content of mature mast cells in the mouse (9,26,29,32-34,41, 48,50,52,57-61), rat (4,10,14,30), dog (7,66), and human (8,19,35,36,51,56,64). Mouse mast cells express varied combinations of a leucine aminopeptidase, a carboxypeptidase [designated mouse mast cell carboxypeptidase A (mMC-CPA)], at least five serine proteases with predicted chymase activity [designated mouse mast cell protease (mMCP) 1, mMCP-2, mMCP-3, mMCP-4, and mMCP-5], and at least two serine proteases with predicted tryptase activity (mMCP-6 and mMCP-7). The cloning of the cDNAs and genes that encode these neutral proteases and the development of mMCP- and mMC-CPA-specific antibodies have provided investigators with the reagents that have enabled them to study the mouse mast cell at a high level of sophistication. This article focuses on the molecular characterization of mouse mast cell proteases, their expression in different populations of mouse mast cells, and the cytokines that control their expression.

CHARACTERIZATION OF MOUSE MAST CELL PROTEASES

Analysis of Mast Cell Protease Genes and Transcripts

mMCP-1 was the first protease to be identified, purified, and to have its amino acid sequence determined (29). Except for mMCP-7 (34), the other proteases were first identified in the whole cell lysates of Kirsten sarcoma virus-immortalized mast cells (KiSV-MC) and BALB/c mouse serosal mast cells (52). Following the separation of their granule proteins by SDS-PAGE, their N-terminal amino acid sequences were determined. Based on the obtained data, oligonucleotides of minimal redundancy were synthesized and then used to screen relevant DNA libraries to obtain the cDNAs and then the genes that encode all but mMCP-3 and the leucine aminopeptidase. All mMCP genes are less than 4 kb in size and possess either 5 or 6 exons. While the mMC-CPA gene has not been sequenced, the human MC-CPA gene is ~32 kb in size and consists of 11 exons. At the nucleotide level, the mMCP-6 (48) and mMCP-7 (34) genes are quite homologous to one another, and therefore these two serine protease genes represent one family of mast cell serine proteases. The mMCP-5 gene is the least homologous in the family but the genes that encode mMCP-1, mMCP-2, mMCP-4, and mMCP-5 (21,25,32,59,60) represent the second family of serine proteases. Chromosomal mapping studies have revealed that the genes that encode mMCP-1, mMCP-2, mMCP-4, and mMCP-5 all reside at a complex on mouse chromosome 14 (21) that also includes the genes that encode cathepsin G, and granzymes B, C, E, and F (5,11,24). As the N-terminal amino acid sequence of mMCP-3 is nearly identical to that of mMCP-4 it is likely that its gene is also located at the chromosome 14 complex. Preliminary mapping studies have revealed that a new serine protease gene resides back to back ~7 kb away from the mMCP-1 gene (Hunt and Stevens, unpublished finding). As the nucleotide sequence of this serine protease gene is quite homologous to that of the mMCP-1 gene, mouse mast cells might express proteases whose cDNAs have not yet been cloned.

While the significance of mMCP gene clustering on chromosomes remains to be determined, it might be important for transcriptional regulation. Experiments are already underway to create mMCP null mice. Since the mMCP-6 and mMCP-7 genes reside on chromosome 17 and the mMC-CPA gene resides on chromosome 3 (21 and Gurish unpublished findings), a classical mating approach probably can be used to eventually obtain a mouse that fails to express a mast cell tryptase, chymase, and mMC-CPA. However, the discovery that the mMCP-1, mMCP-2, mMCP-4, and mMCP-5 genes are very close together on chromosome 14 indicates that a classical mating approach probably can not be used to obtain mice that possess more than one disrupted chymase gene.

The exon/intron organization of the genes that encode the mouse mast cell chymases are highly conserved. All possess 5 exons and the size of these exons are similar among the different mMCPs. The 5' flanking region of the mMCP-2 gene is 89, 93, and 42% identical to that of the 5' flanking regions of the mMCP-1, mMCP-4, and mMCP-5 genes, respectively (21,25,32,60). That the 5' flanking regions of the mast cell chymase genes are more conserved than any of the exons suggests the presence of conserved *cis*-acting elements that regulate their transcription. The 5' flanking region of the mMC-CPA gene possesses a critical GATA-binding motif 51 nucleotides upstream of its transcription-initiation site (51,69). Interestingly, a GATA-binding motif has been detected at a similar location in the 5' flanking regions of the mMCP-1, mMCP-2, and mMCP-4 genes (21). Because mMC-CPA (50) is coordinately expressed with mMCP-4 in serosal mast cells (52) and ear mast cells (61), it is possible that the GATA family of DNA-binding proteins also regulate the transcription of this chymase gene.

Differences have been found in the exon/intron organization of the two mouse mast cell tryptase genes. While the mMCP-6 gene contains 6 exons (48), the mMCP-7 gene contains 5 exons (34). A single nucleotide change in the 3' acceptor site of intron 1 of the mMCP-7 gene prevents this intron from being spliced out during transcription. The result is a 5' untranslated region of 195 nucleotides for the mMCP-7 transcript, as compared to 23 nucleotides for the mMCP-6 transcript.

Analysis of the Deduced Amino Acid Sequences of Mast Cell Proteases

All mMCPs possess the His/Asp/Ser charge-relay amino acids that are found in the active sites of other serine proteases. The specific substrates for these enzymes remain to be determined, but based on *in vitro* studies, mast cell proteases of different species can degrade high molecular weight proteins found in the extracellular matrix such as fibronectin (12,67) and can degrade low molecular weight cell regulatory factors such as angiotensin I (68) and vasointestinal peptide (6). Using pancreatic chymotrypsin as a model (23), the substrate-binding clefts of the chymase mMCPs are formed by residues 176, 197, 198, 199, and 207 (32). No two mMCPs possess the identical amino acids in their substrate-binding clefts. Thus, it is likely that each has a preferred protein substrate. Three-dimensional modeling studies (53) have revealed that the mouse mast cell chymases all possess a peptide loop that extends into the active site, like rat mast cell protease II (47). This loop is not present in pancreatic chymotrypsin (63). Thus, each mouse mast cell serine protease may have a limited substrate specificity. Residue 176 is a Ser in pancreatic chymotrypsin, mMCP-2, and mMCP-4; it is a Thr in mMCP-1. Thus, it is likely that mMCP-1, mMCP-2, and mMCP-4 all possess chymotryptic specificity. However, because mMCP-5 possesses an Asn at residue 176, its substrate specificity might be quite distinct from the other serine proteases whose genes reside at chromosome 14. Like pancreatic trypsin, mMCP-6 and mMCP-7 both possess an Asp at residue 188. Thus, it is likely that both possess tryptic enzymatic activity.

Based on the deduced amino acid sequences of their cDNAs, there is a high degree of similarity among each family member of mMCPs. Whereas mature mMCP-6 is < 30% identical with any mast cell chymase, it is 71% identical with mMCP-7. One area that is highly conserved is the N-terminus. All mouse mast cell chymases possess a Pro-His-Ser-Arg-Pro-Tyr-Met-Ala sequence in their N-terminal regions. Each mouse mast cell protease is initially translated as a zymogen, possessing a hydrophobic signal peptide and an activation peptide. Analogous to other secretory proteins, it is presumed that the signal peptides are removed quite early in the endoplasmic reticulum. It is not known where the activation peptides are removed but only mature proteases have so far been detected in the secretory granules of mouse mast cells (29,50,52).

The hydrophobic signal peptides and the activation peptides are conserved within each family of mMCPs. The mast cell chymases have an 18- or 19-amino acid signal peptide and an activation peptide that is composed of two amino acids. Three of the chymases, mMCP-1, mMCP-2, and mMCP-4, have a Glu-Glu activation peptide and an 18-amino acid signal peptide; mMCP-5 has a Gly-Glu activation peptide and a 19-amino acid signal peptide. The two mouse mast cell tryptases, mMCP-6 and mMCP-7, possess a 10-amino acid activation peptide and an 18-amino acid signal peptide. The last three amino acids of the activation peptide for mMCP-6, Arg-Val-Gly, are identical to those found in other non mast-cell derived tryptases. mMCP-7 has the sequence Arg-Glu-Gly. This suggests that some of the mast cell proteases have a common mechanism of activation.

Granule Packaging of Mast Cell Proteases

Each mast cell protease is stored in the secretory granule in an enzymatically active state. Although the concentration of proteases in the granule is extremely high, they are not rapidly catabolized because they are bound to serglycin proteoglycans. The importance of this mechanism of protease packaging is underscored by the fact that the serglycin gene is expressed before any of the mast cell proteases (17,20). All mouse serglycin proteoglycans contain a novel peptide core that contains a glycosaminoglycan attachment region, possessing alternating Ser and Gly (2,3). Depending on the subclass of mast cells, either heparin or highly sulfated chondroitins are covalently attached to the Ser residues

(45). Based on protein modeling studies, clusters of positively charged amino acids have been identified on the surfaces of rat MC-CPA (10), the mouse mast cell tryptases (26), and the mouse mast cell chymases (53). These positively charged domains probably bind ionically to the negatively charged residues within the glycosaminoglycans of serglycin proteoglycans.

MAST CELL PROTEASE HETEROGENEITY IN THE MOUSE

The development of gene-specific probes and protease-specific antibodies (15,16,22,33) has allowed investigators to phenotype the mast cells that reside at different tissue locations. While all tissue mast cells are derived from a common hematopoietic progenitor cell (40), the mast cells that transiently increase in number in the intestines of helminth-infected BALB/c mice preferentially express mMCP-1 and mMCP-2 (18,29,41,59). In contrast, the mast cells that reside in the serosal cavity of the BALB/c mouse preferentially express mMCP-4, mMCP-5, mMCP-6, and mMC-CPA (32,33,48,50,52,60). While histochemically identical to serosal mast cells, the mast cells that reside in the skin and ears of the BALB/c mouse additionally express mMCP-7 (15,61). This latter finding indicates that mouse mast cells can no longer be grouped into two subclasses. Recently, a strain-dependent expression of mast cell proteases has been observed. The serosal, ear, and skin mast cells of the $WBB6F_1$-+/+ mouse all express mMCP-2, and thus differ from those in the BALB/c mouse (61). The C57BL/6 mouse differs from all other strains in that its ear and skin mast cells do not express mMCP-7 (15). The genetic basis for strain-dependent expression of proteases remains to be determined, but these findings highlight the complexity of mast cell heterogeneity. It has been assumed that the mast cells that reside at a particular site in the human exhibit a specific granule phenotype. Since genetic differences may also occur in the human, this assumption may no longer be correct. The functional significance of strain-dependent expression of proteases in tissues remains to be determined. However, it is of interest to note that the BALB/c mouse differs considerably from the C57BL/6 mouse in its susceptibility to infection and in its immunologic response to a variety of inflammatory agents (31,37).

USE OF *IN VITRO*-DERIVED MOUSE MAST CELLS TO STUDY MAST CELL DIFFERENTION AND MATURATION *IN VITRO* AND *IN VIVO*

Probably the most significant advance in the mouse mast cell field has been the development of different *in vitro* techniques to obtain non-transformed and transformed mast cells. Transformed mouse, rat, and dog mast cells were used to determine the N-terminal sequences of their granule proteases and cell surface receptors. The isolation of sufficient quantities of mRNA from these pure populations of mast cells enabled the construction of mast cell-specific cDNA libraries which in turn led to the cloning and sequencing of the cDNAs that encode these proteases. The ability to transiently transfect immortalized mouse and rat mast cell lines with DNA constructs that possess a reporter gene allowed investigators to identify the *cis*-acting elements in the genes and the *trans*-acting factors in the mast cells that induce transcription of mast cell-specific genes (1,54,69). The non-transformed populations of mast cells have been found to be particularly useful for investigating the effects of different combinations of cytokines on mast cell growth, differentiation, and maturation (13,16-18,20).

In 1981, a number of laboratories discovered that an immature population of non-transformed mast cells developed when mouse bone marrow cells were cultured in the presence of T cell conditioned media (43,55,62). The principal factor in this conditioned media that induces mast cell growth and differentiation was subsequently found to be interleukin (IL) 3 (44). IL-3-developed, BALB/c mouse bone marrow-derived mast cells (mBMMC) contain high steady-state levels of the transcripts that encode mMCP-5, mMCP-6, mMCP-7, and mMC-CPA (32,34,48,50). BALB/c mBMMC that have been developed with c-*kit* ligand (KL) or mBMMC that have been developed with IL-3 followed by KL contain high steady-state levels of the mMCP-4, mMCP-5, mMCP-6, mMCP-7, and mMC-CPA transcripts but not the mMCP-1 or mMCP-2 transcripts (20). While the KL-treated mBMMC exhibit the protease phenotype of a skin mast cell, they remain immature in terms of their morphology and content of granule mediators. If BALB/c mBMMC are cultured in the presence of IL-9 (13) or IL-10 (16-18), they express high steady-state levels of the mMCP-1 and mMCP-2 transcripts, but unlike BALB/c mucosal mast cells, they also contain high steady-state levels of all other protease transcripts. No mast cell has been found in the BALB/c mouse that expresses every mMCP. However, the relevance of these IL-9 and IL-10 findings have become apparent with the discovery, described below, that the mast cells that first develop in the liver and spleen of the V3 mastocytosis mouse also express every mMCP.

The observation that IL-3 and IL-4 both suppress the differentiation effects of IL-9, IL-10, and KL in BALB/c mBMMC (13,17,20) enabled *in vitro* studies to be carried out to assess whether or not the granule phenotype of a non-transformed mast cell could be reversibly altered. Ghildyal and coworkers reported that when IL-10-treated BALB/c mBMMC are subsequently cultured in the presence of IL-3 alone for 1 to 14 days, the steady-state level of the mMCP-2 transcript falls considerably more quickly than the level of mMCP-2 protein (16). Not only did this observation indicate that a mast cell could reversibly alter its granule phenotype but it pointed out the importance of monitoring both mRNA and protein levels when assessing the phenotype of a mast cell at a diseased site.

Mouse mast cell-committed progenitor cells have been found to be particularly susceptible to retroviral transformation. Immortalized mast cell lines have been obtained using Kirsten sarcoma virus (49), Harvey sarcoma virus (46), and Abelson murine leukemia virus (42). While these transformed, immortalized mast cells have been invaluable for the identification, isolation, and characterization of mast cell-specific proteins and genes, their chronic production and release of infectious retrovirus particles has prevented their use *in vivo*. Recently, an immortalized mast cell line (designated V3) was derived by transducing bone marrow cells with pGD$^{v\text{-}abl}$ (Pear, Scott, Austen, Stevens, Webster, Ghildyal, Gurish, and Friend, unpublished finding). This cell does not release retrovirus. Cultured V3 cells express mMCP-5, mMCP-6, and mMC-CPA, but not mMCP-1, mMCP-2, mMCP-4, or mMCP-7. When V3 cells are injected into the tail vein of a BALB/c mouse, an aggressive mastocytosis quickly develops in the liver and spleen, and then more slowly in the stomach, intestine, and lung. The V3 cells that populate the liver differentiate and mature, expressing every mast cell protease so far cloned. In contrast, the V3 cells that populate the intestine preferentially express mMCP-1 and mMCP-2. The ability to alter the granule phenotype of the V3 cell *in vivo* indicates that the V3 mastocytosis mouse represents an excellent *in vivo* system for investigating tissue-specific regulation of mast cell differentiation and maturation.

SUMMARY

Mouse mast cells produce an impressive array of proteases that are all enzymatically active at neutral pH. While initially translated as zymogens, each is packaged in the secretory granule in an enzymatically active state. Because different mouse mast cells express varied combinations of mMC-CPA and mMCP-1 to mMCP-7, these granule constituents have been used as markers to study the development of hematopoietic progenitor cells into mature mast cells. Based on *in vitro* and *in vivo* studies, the granule phenotype of a mouse mast cell is determined, in part, by what cytokines the mast cell is exposed to in its microenvironment.

ACKNOWLEDGMENT

JEH is the recipient of a C.J. Martin Research Fellowship from the NH & MRC of Australia.

REFERENCES

1. Avraham S, Avraham H, Austen KF, Stevens RL. Negative and positive *cis*-acting regulatory elements in the 5' flanking region of the mouse gene that encodes the serine/glycine-rich peptide core of secretory granule

proteoglycans. *J Biol Chem* 1992;267:610-7.
2. Avraham S, Austen KF, Nicodemus CF, Gartner MC, Stevens RL. Cloning and characterization of the mouse gene that encodes the peptide core of secretory granule proteoglycans and expression of this gene in transfected rat-1 fibroblasts. *J Biol Chem* 1989;264:16719-26.
3. Avraham S, Stevens RL, Nicodemus CF, Gartner MC, Austen KF, Weis JH. Molecular cloning of a cDNA that encodes the peptide core of a mouse mast cell secretory granule proteoglycan and comparison with the analogous rat and human cDNA. *Proc Natl Acad Sci USA* 1989;86:3763-7.
4. Benfey PN, Yin FH, Leder P. Cloning of the mast cell protease, RMCPII. Evidence for cell-specific expression and a multi-gene family. *J Biol Chem* 1987;262:5377-84.
5. Brunet JF, Dosseto M, Denizot F, et al. The inducible cytotoxic T-lymphocyte-associated gene transcript CTLA-1 sequence and gene localization to mouse chromosome 14. *Nature* 1986;322:268-71.
6. Caughey GH. Roles of mast cell tryptase and chymase in airway function. *Am J Physiol* 1989;257:L39-46.
7. Caughey GH, Raymond WW, Vanderslice P. Dog mast cell chymase: molecular cloning and characterization. *Biochemistry* 1990;29:5166-71.
8. Caughey GH, Zerwick EH, Vanderslice P. Structure, chromosome assignment, and deduced amino acid sequence of a human gene for mast cell chymase. *J Biol Chem* 1991;266:12956-63.
9. Chu W, Johnson DA, Musich PR. Molecular cloning and characterization of mouse mast cell chymases. *Biochim Biophy Acta* 1992;1121:83-7.
10. Cole KR, Kumar S, Le Trong H, Woodbury RG, Walsh KA, Neurath H. Rat mast cell carboxypeptidase: Amino acid sequence and evidence of enzymatic activity within mast cell granules. *Biochemistry* 1991;30:648-55.
11. Crosby JL, Bleackley RC, Nadeau JH. A complex of serine protease genes expressed preferentially in cytotoxic T-lymphocytes is closely linked to the T-cell receptor α- and δ-chain genes on mouse chromosome 14. *Genomics* 1990;6:252-9.
12. DuBuske L, Austin KF, Czop J, Stevens RL. Granule-associated serine neutral proteases of the mouse bone marrow-derived mast cell that degrade fibronectin: Their increase after sodium butyrate treatment of the cells. *J Immunol* 1984;133:1535-41.
13. Eklund KK, Ghildyal N, Austen KF, Stevens RL. Induction by IL-9 and suppression by IL-3 and IL-4 of the levels of chromosome 14-derived transcripts that encode late-expressed mouse mast cell proteases. *J Immunol* 1993;151:4266-73.
14. Everitt MT, Neurath H. Rat peritoneal mast cell carboxypeptidase: localization, purification, and enzymatic properties. *FEBS Letters* 1980;110:292-6.
15. Ghildyal N, Friend DS, Freelund R, et al. Lack of expression of the tryptase, mMCP-7, in C57BL/6J mouse mast cells. *J Immunol*, in press.

16. Ghildyal N, Friend DS, Nicodemus CF, Austen KF, Stevens RL. Reversible expression of mouse mast cell protease-2 mRNA and protein in cultured mast cells exposed to IL-10. *J Immunol* 1993;151:3206-14.
17. Ghildyal N, McNeil HP, Gurish MF, Austen KF, Stevens RL. Transcriptional regulation of the mucosal mast cell-specific protease gene, mMCP-2, by interleukin 10 and interleukin 3. *J Biol Chem* 1992;267:8473-7.
18. Ghildyal N, McNeil HP, Stechschulte S, et al. IL-10 induces transcription of the gene for mouse mast cell protease-1, a serine protease preferentially expressed in mucosal mast cells of *Trichinella spiralis*-infected mice. *J Immunol* 1992;149:2123-9.
19. Goldstein SM, Kaempfer CE, Proud D, Schwartz LB, Irani A-M, Wintroub BU. Detection and partial characterization of a human mast cell carboxypeptidase. *J Immunol* 1987;139:2724-9.
20. Gurish MF, Ghildyal N, McNeil HP, Austen KF, Gillis S, Stevens RL. Differential expression of secretory granule proteases in mouse mast cells exposed to interleukin 3 and c-*kit* ligand. *J Exp Med* 1992;175:1003-12.
21. Gurish MF, Nadeau JH, Johnson KR, et al. A closely linked complex of mouse mast cell-specific chymase genes on chromosome 14. *J Biol Chem* 1993;268:11372-9.
22. Gurish MF, Nicodemus CF, Ghildyal N, Austen KF, Stevens RL. Anti-peptide antibodies against the exopeptidase, mouse mast cell carboxypeptidase A (mMC-CPA). *J Allergy Clin Immunol* 1993;89:309 (abstract).
23. Hartley BS. Homologies in serine proteases. *Philos Trans R Soc Lond B Biol Sci* 1970;257:77-87.
24. Heusel JW, Scarpati EM, Jenkins NA, et al. Molecular cloning, chromosomal location, and tissue-specific expression of the murine cathepsin G gene. *Blood* 1993;81:1614-23.
25. Huang R, Blom T, Hellman L. Cloning and structural analysis of mMCP-1, mMCP-4 and mMCP-5, three mouse mast cell-specific serine proteases. *Eur J Immunol* 1991;21:1611-21.
26. Johnson DA, Barton GJ. Mast cell tryptases: Examination of unusual characteristics by multiple sequence alignment and molecular modeling. *Protein Sci* 1992;1:370-7.
27. Kitamura Y, Go S, Hatanaka K. Decrease of mast cells in W/W^v mice and their increase by bone marrow transplantation. *Blood* 1978;52:447-52.
28. Kitamura Y, Go S. Decreased production of mast cells in Sl/Sl^d anemic mice. *Blood* 1979; 53:492-7.
29. Le Trong H, Newlands GFJ, Miller HRP, Charbonneau H, Neurath H, Woodbury RG. Amino acid sequence of a mouse mucosal mast cell protease. *Biochem*istry 1989;28:391-5.
30. Le Trong H, Parmelee DC, Walsh KA, Neurath H, Woodbury RG. Amino acid sequence of rat mast cell protease I (chymase). *Biochemistry*

1987;26:6988-94.
31. Levitt RC, Mitzner W. Autosomal recessive inheritance of airway hyperreactivity to 5-hydroxytryptamine. *J Appl Physiol* 1989;67:1125-32.
32. McNeil HP, Austen KF, Somerville LL, Gurish MF, Stevens RL. Molecular cloning of the mouse mast cell protease-5 gene. A novel secretory granule protease expressed early in the differentiation of serosal mast cells. *J Biol Chem* 1991;266:20316-22.
33. McNeil HP, Frenkel DP, Austen KF, Friend DS, Stevens RL. Translation and granule localization of mouse mast cell protease-5. *J Immunol* 1992;149:2466-72.
34. McNeil HP, Reynolds DS, Schiller V, et al. Isolation, characterization, and transcription of the gene encoding mouse mast cell protease 7. *Proc Acad Natl Sci USA* 1992;89:11174-8.
35. Miller JS, Moxley G, Schwartz LB. Cloning and characterization of a second complementary DNA for human tryptase. *J Clin Invest* 1990;86:864-70.
36. Miller JS, Westin EH, Schwartz LB. Cloning and characterization of complementary DNA for human tryptase. *J Clin Invest* 1989;84:1188-95
37. Mitchell G F, Hogarth-Scott RS, Edwards RD, Lewers HM, Cousins G, Moore T. Studies on immune responses to parasite antigens in mice. *Int Archs Allergy Appl Immunol* 1976;52:64-.
38. Nabel G, Galli SJ, Dvorak AM, Dvorak HF, Cantor H. Inducer T lymphocytes synthesize a factor that stimulates proliferation of cloned mast cells. *Nature* 1981;291:332-4.
39. Nagao K, Yokoro K, Aaronson SA. Continuous lines of basophil/mast cells derived from normal mouse bone marrow. *Science* 1981;212:333-5.
40. Nakano T, Sonada T, Hayashi C, et al. Fate of bone marrow-derived cultured mast cells after intracutaneous, intraperitoneal, and intravenous transfer into genetically mast cell deficient W/W^v mice. *J Exp Med* 1985;162:1025-43.
41. Newlands GF, Gibson S, Knox DP, Grencis R, Wakelin D, Miller HR. Characterization and mast cell origin of a chymotrypsin-like proteinase isolated from intestines of mice infected with *Trichinella spiralis*. *Immunology* 1987;62:629-34.
42. Pierce JH, Di Fiore PP, Aaronson SA, et al. Neoplastic transformation of mast cells by Abelson-MuLV: Abrogation of IL-3 dependence by a nonautocrine mechanism. *Cell* 1985;41:685-93.
43. Razin E, Cordon-Cardo C, Good RA. Growth of a pure population of mouse mast cells in vitro with conditioned medium derived from concanavalin-A stimulated splenocytes. *Proc Natl Acad Sci USA* 1981;78:2559-61.
44. Razin E, Ihle JN, Seldin D, Mencia-Huerta J-M, et al. Interleukin 3: A differentiation and growth factor for the mouse mast cell that contains chondroitin sulfate E proteoglycan. *J Immunol* 1984;132:1479-86.

45. Razin E, Stevens RL, Akiyama F, Schmid K, Austen KF. Culture from mouse bone marrow of a subclass of mast cells possessing a distinctive chondroitin sulfate proteoglycan with glycosaminoglycans rich in N-acetylgalactosamine-4,6-disulfate. *J Biol Chem* 1982;257:7229-36.
46. Rein A, Keller J, Schultz AM, Holmes KL, Medicus R, Ihle JN. Infection of immune mast cells by Harvey sarcoma virus: immortalization without loss of requirement for interleukin-3. *Molec Cell Biol* 1985;5:2257-64.
47. Remington SJ, Woodbury RG, Reynolds RA, Matthews BW, Neurath H. The structure of rat mast cell protease II at 1.9-Å resolution. *Biochemistry* 1988; 27:8097-105.
48. Reynolds DS, Gurley, DS, Austen KF, Serafin WE. Cloning of the cDNA and gene of mouse mast cell protease-6. Transcription by progenitor mast cells and mast cells of the connective tissue subclass. *J Biol Chem* 1991;266;3847-53.
49. Reynolds DS, Serafin WE, Faller DV, et al. Immortalization of murine connective tissue-type mast cells at multiple stages of their differentiation by coculture of splenocytes with fibroblasts that produce Kirsten sarcoma virus. *J Biol Chem* 1988;263:12783-91.
50. Reynolds DS, Stevens RL, Gurley DS, Lane WS, Austen KF, Serafin WE. Isolation and molecular cloning of mast cell carboxypeptidase A: A novel member of the carboxypeptidase gene family. *J Biol Chem* 1989;264:20094-9.
51. Reynolds DS, Gurley DS, Stevens RL, Sugarbaker DJ, Austen KF, Serafin WE. Cloning of cDNAs that encode human mast cell carboxypeptidase A, and comparison of the protein with mouse mast cell carboxypeptidase A and rat pancreatic carboxypeptidases. *Proc Natl Acad Sci USA* 1989;86:9480-4.
52. Reynolds DS, Stevens RL, Lane WS, Carr MH, Austen KF, Serafin WE. Different mouse mast cell populations express various combinations of at least six distinct mast cell serine proteases. *Proc Natl Acad Sci USA* 1990;87:3230-4.
53. Šali A, Matsumoto R, McNeil HP, Karplus M, Stevens RL. Three dimensional models of four mast cell chymases. *J Biol Chem* 1993; 268:9023-34.
54. Sarid J, Benfey PN, Leder P. The mast cell-specific expression of a protease gene, RMCP-II, is regulated by an enhancer element that binds specifically to mast cell *trans*-acting factors. *J Biol Chem* 1989;264:1022-6.
55. Schrader JW, Lewis SJ, Clark-Lewis I, Culvenor JG. The persisting (P) cell: Histamine content, regulation by a T cell-derived factor, origin from a bone marrow precursor, and relationship to mast cells. *Proc Natl Acad Sci USA* 1981;78:323-7.
56. Schwartz LB, Lewis RA, Austin KF. Tryptase from human pulmonary mast cells: purification and characterization. *J Biol Chem* 1981; 256:11939-43.
57. Serafin WE, Dayton ET, Gravallese PM, Austen KF, Stevens RL.

Carboxypeptidase A in mouse mast cells: Identification, characterization, and use as a differentiation marker. *J Immunol* 1987; 139:3771-6.
58. Serafin WE, Guidry UA, Dayton ET, Kamada MM, Stevens RL, Austen KF. Identification of aminopeptidase activity in the secretory granules of mouse mast cells. *Proc Natl Acad Sci USA* 1991;88:5984-8.
59. Serafin WE, Reynolds DS, Rogelj S, et al. Identification and molecular cloning of a novel mouse mucosal mast cell serine protease. *J Biol Chem* 1990;265:423-9.
60. Serafin, WE, Sullivan TP, Conder GA, et al. Cloning of the cDNA and gene for mouse mast cell protease 4. Demonstration of its late transcription in mast cell subclasses and analysis of its homology to subclass-specific neutral proteases of the mouse and rat. *J Biol Chem* 1991;266:1934-41.
61. Stevens RL, Friend DS, McNeil HP, Schiller V, Ghildyal N, Austen KF. Strain-specific and tissue-specific expression of mouse mast cell secretory granule proteases. *Proc Natl Acad Sci USA* 1994;91:128-32.
62. Tertian G, Yung Y-P, Guy-Grand D, Moore MAS. Long-term *in vitro* culture of murine mast cells. I. Description of a growth factor-dependent culture technique. *J Immunol* 1981;127:788-94.
63. Tsukada H, Blow DM. Structure of α-chymotrypsin refined at 1.68 Å resolution. *J Mol Biol* 1985; 184:703-11.
64. Urata H, Kinoshita A, Perez DM, et al. Cloning of the gene and cDNA for human heart chymase. *J Biol Chem* 1991;266:17173-9.
65. Vanderslice P, Ballinger SM, Tam EK, Goldstein SM, Craik CS, Caughey GH. Human mast cell tryptase: multiple cDNAs and genes reveal a multigene serine protease family. *Proc Natl Acad Sci USA* 1990;87:3811-5.
66. Vanderslice P, Craik CS, Nadel JA, Caughey GH. Molecular cloning of dog mast cell tryptase and a related protease: structural evidence of a unique mode of serine protease activation. *Biochemistry* 1989;28:4148-55.
67. Vartio T, Sëppa H, Vaheri A. Susceptibility of soluble and matrix fibronectins to degradation by tissue proteinases, mast cell chymase and cathepsin G. *J Biol Chem* 1981;256:471-7.
68. Wintraub BU, Schechter NM, Lazarus SC, Kaempfer CE, Schwartz LB. Angiotensin-I conversion by human and rat chymotryptic proteinases. *J Invest Dermatol* 1984;83:336-9.
69. Zon LI, Gurish MF, Stevens RL, et al. GATA-binding transcription factors in mast cells regulate the promoter of the mast cell carboxypeptidase A gene. *J Biol Chem* 1991;266:22948-53.

14

STRUCTURE AND FUNCTION OF HUMAN MAST CELL TRYPTASE

Lawrence B. Schwartz

Department of Internal Medicine, Virginia Commonwealth University, Richmond, VA 23298

Mast cells and basophils are the major effector cells in human allergic diseases. Mast cells, when activated by either IgE-dependent or IgE-independent agonists, secrete an array of mediators that act on the vasculature, smooth muscle, connective tissue, mucous glands and inflammatory cells to cause immediate-type hypersensitivity reactions. Although their immunopathological role in these reactions is acknowledged, their purpose in normal tissue homeostasis or in defense against microorganisms each remain less certain. Preformed mediators, stored in secretory granules and secreted with cell activation, include a biogenic amine, typically histamine, proteoglycans, either heparin, over-sulfated chondroitin sulfates or both, and a spectrum of neutral proteases.

Neutral proteases, accounting for the vast majority of the granule protein, serve as selective markers of mast cells and of different types of mast cells (1). Tryptase is the principal enzyme accounting for the trypsin-like activity first detected in human mast cells by histochemical techniques (2,3). Tryptase, along with chymase, carboxypeptidase and a cathepsin G-like protease are the dominant protein components in secretory granules of the MC_{TC} type of human mast cell, whereas only tryptase is detected in those of the MC_T type of mast cell. Whereas human mast cells are rich in serine-type proteases, human basophils exhibit relatively little protease activity (4). Because of the selective expression of these proteases in mast cells, one could hypothesize that these enzymes play an important role in the biology of this cell type, that such enzymes might serve as sensitive and specific markers of mast cells and of mast cell activation, and that regulation of the transcription of their corresponding genes is related to the regulation of the differentiation of this cell type. This chapter will focus on studies relating to human mast cells.

PHYSICAL AND ENZYMATIC PROPERTIES

Tryptase was first purified to apparent homogeneity from dispersed and enriched lung mast cells in 1981 (2), and later from crude lung cell dispersions (5), HMC-1 cells (6) and from lung (7), pituitary (8), and skin (9) tissue by conventional chromatography, and more recently from lung by immunoaffinity chromatography (10). Tryptase is a tetrameric endoprotease of 134,000 daltons with subunits of 31,000 to 34,000 daltons, each with an active enzymatic site (2), specificity for basic amino acid residues (11) and common antigenic sites (12). A reduction in mw of tryptase subunits of 2,000 to 4,000 after treatment with endoglycosidase indicates that carbohydrate is present on each subunit (8). Within mast cell granules and after secretion tryptase associates with proteoglycan, presumably heparin, by ionic interactions. Because NaCl concentrations of at least 0.7 M are needed to dissociate tryptase from heparin, the mechanism for dissociation to occur *in vivo*, if one exists at all, is not clearly defined. However, antithrombin III, when present in large excess *in vitro*, can dissociate up to one-third of the tryptase bound to heparin. Chymase and carboxypeptidase appear to reside in a complex with proteoglycan that is distinct and separable from the macromolecular tryptase-proteoglycan complex (13), suggesting that tryptase is processed and packaged separately from these other two proteases.

Although classified as a serine protease, the absence of inhibition by the classical inhibitors of serine esterases present in plasma, lung and urine as well as by lima bean, soy bean and ovomucoid trypsin inhibitors clearly distinguish tryptase from pancreatic trypsin and from most other serine esterases (14). Tryptase is inhibited by small molecular wight substances such as leupeptin, diisopropyl fluorophosphate, and phenyl methyl sulfonyl fluoride. Divalent cations like calcium (15), and benzamidine and its derivatives (16,17) are competitive inhibitors of tryptase. Interestingly, histamine shifts the substrate dose-response curve to the right and to a sigmoidal pattern, suggesting cooperative behavior (15).

Instead of the enzymatic activity of tryptase being regulated by the usual inhibitors of serine proteases, tryptase is uniquely stabilized in its active tetrameric form by heparin (5), to which it is ionically bound under physiologic conditions. Negative charge density rather than carbohydrate structure is the primary determinant for stabilizing tryptase activity (18). When free in solution, tryptase subunits irreversibly dissociate from one another into inactive monomers, without any evidence for autodegradation. Substantial conformational changes occur during this process as evidenced by circular dichroic spectral shifts and by distinct epitopes being detected on the active tetramer and inactive monomers (10). It also appears that inactivation proceeds as a multistep process, because limited recovery of lost enzyme activity reportedly can be recovered at an early stage of the inactivation process (19). The physiologic means of dissociating tryptase from heparin or chondroitin sulfate E *in vivo* is not totally clear, but the capacity of divalent cations (15) and of heparin binding proteins such as antithrombin III (14) to destabilize heparin-bound tryptase at physiologic concentrations *in vitro* may be relevant to the situation *in vivo*.

BIOLOGICAL ACTIVITY

Even though the biologic role of human tryptase *in vivo* has not been convincingly demonstrated, several activities of potential biologic interest have been examined *in vitro* (Table 1). Tryptase rapidly cleaves and inactivates fibrinogen as a coagulable substrate for thrombin (20); the lack of fibrin deposition and rapid resolution of urticaria/angioedema reactions may, in part, reflect this same activity *in vivo*. Tryptase directly degrades fibronectin (21). Tryptase activates latent collagenase derived from rheumatoid synovial cells, apparently by first activating prostromelysin (metalloproteinase III) (22), which in turn activates latent collagenase. The increased numbers of mast cells found in rheumatoid synovium (23) and in inflammatory cutaneous lesions of scleroderma (24) together with the increased tissue turnover that occurs in each condition suggest a related role for tryptase. Tryptase does not appear to affect complement anaphylatoxin or bradykinin generation (25); an earlier report of C3a generation from C3 could not be reproduced with immunoaffinity-purified tryptase (Schwartz, unpublished data). Bradykinin, lysylbradykinin, C3a and C5a, if generated *in vivo* during immediate hypersensitivity events, is not directly due to tryptase activity. Tryptase degrades vasoactive intestinal peptide (VIP), peptide histidine-methionine (PHM) and calcitonin gene-related peptide (CGRP) (26,27), neuropeptide that relax smooth muscle. Inreased degradation of such neuropeptides in lung could contribute to airway hyperreactivity in asthma. Tryptase also exhibits mitogenic activity on fibroblast lines *in vitro* (28), suggesting one mechanism by which mast cells might participate in fibrogenesis in diseases such as scleroderma.

Table 1
Potential Biological Activities of Human Tryptase

Enzymatic Event	Potential Significance
Fibrinogenolysis	Prevent clotting at local sites of mast cell activation
Degrades CGRP, VIP & PHM	Decrease bronchodilatory activities of these neuropeptides
Activate prostromelysin	Facilitate activation of procollagenase and tissue remodeling
Degrade fibronectin	Tissue remodeling
Stimulate fibroblast proliferation	Enhance fibrogenesis or wound repair

MOLECULAR BIOLOGY

Two tryptase cDNA molecules (α and ß) have been cloned from human lung mast cell cDNA library and sequenced (29,30), three (I, II and III) from a human skin mast cell library, as summarized in Table 2 (31). The α-tryptase cDNA codes for a 245 amino acid catalytic protein with two putative carbohydrate binding sites (29,32). This cDNA was originally reported as a 244 amino acid protein due to three guanine residues that were undetected due to an unresolved compression. The β-tryptase cDNA codes for a 245 amino acid catalytic protein that is 92% identical to α-tryptase and 98% to 99% identical tryptases I, II and III. There are 2 N-linked carbohydrate binding sites in the α, I and III types of tryptase, and 1 carbohydrate binding site in the other tryptase sequences, suggesting an explanation for a portion of the electrophoretic heterogeneity of natural tryptase in SDS polyacrylamide gels. Although the active form of tryptase seemingly can be composed of homologous gene products, isoforms of tryptase have not been ruled out. In certain species of mice, only MMCP-6 mRNA is detected in connective tissue mast cells, which contain active tryptase. In the human basophilic cell line called KU-812 (33), only tryptase-ß mRNA was detected, and only tryptase-α mRNA was detected in Mono-Mac-6 cells (32), though in each case proteolytic activity was not evaluated.

Table 2
Molecular Characteristics of Human Tryptase

Property	α-Tryptase	β-like Tryptases (I,II,III,β)
Number of amino acids: catalytic regions	245	245
prepro regions	30	30
Catalytic region: α to β homology	92% identical	
β to I,II,III homologies	98% to 100% identical	
Prepro region: α to β homology	87%	
β to I,II,III homologies	100%	
Carbohydrate binding regions	2	1 to 2
Calculated peptide molecular weights		
catalytic region	27,682	27,449 - 27,585
prepro region	3,048	3,089
Chromosome localization	16	16

The prepropeptide portion is 30 amino acids long in all cases; these regions are identical in ß, I, II and III forms of tryptase, but differ by 4 amino acids from the α-tryptase prepropeptide. The activation peptide for tryptase has not been clearly defined, but its termination in Gly is unusual. The propeptide sequences for pancreatic serine proteases terminate in a basic residue; most leukocyte and mast cell granule proteases terminate in an acidic residue. Neither the prepropeptide nor the propeptide portions of tryptase have been detected in enzyme purified from human lung mast cells, consistent with tryptase being stored in an active form. The prepropeptide may be responsible for directing tryptase to secretory granules and/or the assembly of tryptase as an active tetramer, and most likely are removed before or shortly after tryptase enters the secretory granules.

Tryptase, a tetrameric protease in mammals, appears to originate from at least one of two genes, in mice (34,35) and humans (30,31). Sequences for human tryptase-ß and -α each reside in the normal haploid chromosome on chromosome 16. Whether more than two genes for tryptase are present has not been established. The one gene for tryptase in humans that has been studied contains 6 exons and intron splice phases of 0, I, II, I and 0. The first intron falls between the 5'-noncoding and the ATG-methionine start site, which is rare among proteases. Positions of the 4 introns within the coding region are similar to those for trypsin and glandular kallikrein, but different from those for proteases involved in blood coagulation. The 5' region has several consensus sites for regulating transcription, including TATA and CAAT boxes and a region showing homology to the enhancer elements for the rat chymase II gene (36) and for protease genes expressed in the pancreas. If the latter proves to be a mast cell specific enhancer region, it would be of interest to examine its regulation during the differentiation of human mast cells. Each of the cloned tryptases has an activation peptide terminating in Gly, somewhat unique among serine proteases, and suggestive of a common mechanism for activation. The catalytic portions each contain four putative disulfide bonds. Two large insertions have been predicted to reside on either side of the substrate binding cleft, possibly explaining the restricted substrate specificity and resistance to inhibition by biologic inhibitors of serine proteases (37).

CELLULAR MARKER OF MAST CELLS DURING DEVELOPMENT AND IN TISSUES

Neutral proteases, because each represents a single gene product, often produced selectively and in great abundance in mast cells, best reflect the heterogeneity or plasticity of mast cells *in vivo* and *in vitro*, particularly in humans where histochemical heterogeneity is less apparent than in rodents. By immunohistochemical analyses two types of mast cells have been found. The MC_{TC} type contains tryptase, chymase, cathepsin G-like protease and mast cell carboxypeptidase, and predominates in normal skin and intestinal submucosa; the MC_T type contains only tryptase, and predominates in normal intestinal mucosa and lung alveolar wall. Nearly equivalent concentrations of each type are found in nasal mucosa. Substantial amounts of tryptase are present in MC_{TC} cells derived from foreskin (35 pg/cell) and in MC_T cells derived from lung (10 pg/cell), where it is located in secretory granules and is released in parallel with histamine during degranulation (38). These levels appear to account for perhaps 20% of the entire protein in the mast cell. Small amounts have been measured in human basophils (0.04 pg/cell). Other cell types found in peripheral blood such as eosinophils, neutrophils, and lymphocytes, and in lung, skin and bowel have no detectable tryptase protein. Thus the enzyme is a discriminating marker for human mast cells.

In MC_{TC} cells, tryptase, chymase and mast cell carboxypeptidase reside in macromolecular complexes with proteoglycan. Interestingly, tryptase resides in a separate complex from that in which chymase and mast cell carboxypeptidase are found (13), even though all are detected in the same secretory granules. Unlike for rodent mast cells histochemical heterogeneity based on the presence or absence of heparin proteoglycan does not appear to be useful to distinguish different types of human mast cells, because all human mast cells contain heparin based on an ultrastructural analysis using antithrombin III-gold (39).

Because tryptase expression occurs in mast cells at the time granule formation begins (40), the enzyme is a useful marker of differentiation for this cell type. Studies of mast cell differentiation mediated by stem cell factor (Kit-ligand) with progenitors derived from fetal liver cells (41) and from cord blood mononuclear cells (42,43) have utilized tryptase expression as a critical marker for identifying commitment to a mast cell lineage. Divergent or distinct lineages for these mast cell types are suggested by the apparent commitment of immature mast cells to one or the other phenotype as assessed by electron microscopy and immunogold staining of immature mast cells in normal tissues (40) and by the development *in vitro* of mostly MC_{TC} cells from cord blood mononuclear cell progenitors (44) and mostly MC_T cells from fetal liver progenitors (45). Preliminary data also suggest that lung-derived MC_T cells contain no detectable chymase mRNA, and remain as MC_T cells when cultured on mouse 3T3 fibroblasts (Schwartz et al., unpublished results).

A CLINICAL INDICATOR OF THE ACTIVATION AND BURDEN OF MAST CELLS

A clinically significant use for tryptase in humans as a precise marker of mast cell activation, and perhaps of total mast cell burden, has emerged. The advantages of tryptase over other potential indicators are its abundant and selective presence in mast cell secretory granules, and its persistence in the circulation after release. A specific sandwich immunoassay was developed using the mouse mAb called G5 for capture to precisely quantify tryptase in complex biologic fluids (46,47) such as serum (48), bronchoalvoelar lavage fluid (49), nasal lavage fluid (50), skin chamber fluid (51,52) and tears (53). In serum, tryptase levels in normal subjects are undetectable (< 1 ng/ml), whereas elevated levels are detected in systemic mast cell disorders, such as anaphylaxis and mastocytosis (Table 3). Other disorders resulting in various shock syndromes such as septic shock, vasovagal reactions, myocardial shock, toxic shock syndrome (Schwartz, L.B. and Leung, D.Y.M., unpublished data) and the Mazzottii reaction (Schwartz, L.B. and Nutman, T.B., unpublished data) seen with treatment of Onchocerciasis with diethylcarbamazine (54) do not exhibit elevations of serum tryptase, suggesting that mast cell activation is not the predominant pathway involved. Thus, tryptase levels provide a more precise assessment of mast cell activation *in vivo* than previously available.

Table 3
Levels of Tryptase in Serum or Plasma
(G5 capture antibody assay)

Clinical condition	Tryptase (ng/ml)
Normal level	<1
Systemic anaphylaxis (mast cell-dependent)	
Severe within 4 hours of hypotension	>5
Postmortem, fatal anaphylaxis	>10
Moderate within 2 to 4 hours of signs	1 to 5
Severe after 4 hours of hypotension	varies
Systemic mastocytosis	
Non-acute, baseline	≥1 (~50% of subjects)
Acute	>1 (nearly 100% of subjects)
Sudden Infant Death Syndrome	>10 (~40% of victims)
Septic shock, myocardial shock, vasovagal reactions, serum sickness, vasculitis, trauma, toxic shock syndrome, Mazzotti reaction	<1

Tryptase quantities in serum are elevated in systemic anaphylaxis (48). During bee sting-induced anaphylaxis circulating levels are maximal at about 60 minutes after the sting, or 30 to 60 min after the onset of clinical symptoms, and then decline with a half-life of 1.5 to 2.5 hours (55). The magnitude of the tryptase level correlates with the clinical severity as measured by the drop in mean arterial pressure (56). Elevated levels of mast cell tryptase in postmortem sera reflect antemortem mast cell activation and suggest that tryptase levels also may be used as a practical indicator of fatal anaphylaxis (57). Elevated levels of tryptase in postmortem sera of victims of Sudden Infant Death Syndrome (SIDS) compared to controls has implicated mast cell activation in the pathogenesis of this disorder (58), suggesting that anaphylaxis accounts for the pathogenesis of a substantial portion of SIDS victims.

Tryptase levels in serum may reflect the tissue burden of mast cells as well as their activation. In subjects with systemic mastocytosis, characterized by mast cell hyperplasia in skin, liver, spleen and bone marrow, elevated levels of tryptase are found in about 50% of subjects at baseline, i.e., when not acutely symptomatic. Preliminary results with a more sensitive immunoassay for tryptase suggest that elevated levels of the protein will be detectable in nearly all subjects having this disease, making the new assay for tryptase (see below) a useful screening tool.

Ongoing mast cell activation in asthma appears to be a characteristic of this chronic inflammatory disease. It is detected by elevated levels of tryptase and PGD_2 in bronchoalveolar lavage fluid, higher spontaneous release of histamine by mast cells obtained from the bronchoalveolar lavage fluid of asthmatics than nonasthmatics and ultrastructural analysis of mast cells in pulmonary tissue. Both in lung and in skin, tryptase release occurs during the immediate response to allergen, indicating mast cell degranulation, but not during the late response, which instead involves eosinophils and perhaps basophils (59). Mast cell activation occurs during the immediate respiratory and vascular responses to oral aspirin based on elevated levels of serum tryptase (60), but may not be a feature of exercise-induced asthma, because levels of tryptase and other mast cell mediators in bronchoalveolar lavage fluid do not increase with exercise (61).

Using a new mouse mAb for capture called B12, a more sensitive sandwich immunoassay for tryptase was developed (62). Increased sensitivity results both from an apparent increased binding affinity, and detection of a form of tryptase not recognized by G5. Although both assays detect tryptase extracted from mast cells dispersed from lung and skin, in the media or cell extracts of fetal liver mast cells in culture, and in sera obtained during systemic anaphylaxis comparably well, tryptase present in sera of healthy subjects is only detected by the B12-based assay. Normal levels of serum tryptase range from about 1 to 20 ng/ml. In subjects with active allergic rhinitis with high tryptase levels in nasal fluid, serum levels are no different than in controls. Of potential interest is the observation that bee sting volunteers destined to have severe hypotensive reactions have elevated levels of tryptase at baseline compared to sensitive subjects who exhibit only mild reactions. Also, tryptase levels in subjects with systemic mastocytosis during nonacute periods are markedly elevated with the B12-based assay, whereas only about 50% of the samples show modest elevations with the G5-based assay (Schwartz and Metcalfe, unpublished observations). Thus, tryptase measured with the new immunoassay may prove to be a useful measure of mast cell burden, and under certain conditions may predict risk of anaphylaxis.

REFERENCES

1. Neutral proteases of mast cells. In: Schwartz LB, Monographs in Allergy Vol. 27, Basel: Karger, 1990: pp. 1-162.
2. Schwartz LB, Lewis RA, Austen KF. Tryptase from human pulmonary mast cells. Purification and characterization. *J Biol Chem.* 1981;256:11939-43.
3. Hopsu VK, Glenner GG. A histochemical enzyme kinetic system applied to the trypsin-like amidase and esterase activity in human mast cells. *J Cell Biol.* 1963;17:503-10.
4. Yam LT, Li CY, Crosby WH. Cytochemical identification of monocytes and granulocytes. *Am J Clin Pathol.* 1971;55:283-90.

5. Schwartz LB, Bradford TR. Regulation of tryptase from human lung mast cells by heparin. Stabilization of the active tetramer. *J Biol Chem.* 1986;261:7372-9.
6. Butterfield JH, Weiler DA, Hunt LW, Wynn SR, Roche PC. Purification of tryptase from a human mast cell line. *J Leukocyte Biol.* 1990;47:409-19.
7. Smith TJ, Hougland MW, Johnson DA. Human lung tryptase, purification and characterization. *J Biol Chem.* 1984;259:11046-51.
8. Cromlish JA, Siedah NG, Marcinkiewicz M, Hamelin J, Johnson DA, Chretein M. Human pituitary tryptase: molecular forms, NH -terminal sequence, immunocytochemical localization, and specificity with prohormone and fluorogenic substrates. *J Biol Chem.* 1987;262:1363-73.
9. Harvima IT, Schechter NM, Harvima RJ, Fräki JE. Human skin tryptase: Purification, partial characterization and comparison with human lung tryptase. *Biochim Biophys Acta.* 1988;957:71-80.
10. Schwartz LB, Bradford TR, Lee DC, Chlebowski JF. Immunologic and physicochemical evidence for conformational changes occurring on conversion of human mast cell tryptase from active tetramer to inactive monomer: Production of monoclonal antibodies recognizing active tryptase. *J Immunol.* 1990;144:2304-11.
11. Tanaka T, McRae BJ, Cho K, et al. Mammalian tissue trypsin-like enzymes. Comparative reactivities of human skin tryptase, human lung tryptase and bovine trypsin with peptide 4-nitroanilide and thioester substrates. *J Biol Chem.* 1983;258:13552-9.
12. Schwartz LB. Monoclonal antibodies against human mast cell tryptase demonstrate shared antigenic sites on subunits of tryptase and selective localization of the enzyme to mast cells. *J Immunol.* 1985;134:526-31.
13. Goldstein SM, Leong J, Schwartz LB, Cooke D. Protease composition of exocytosed human skin mast cell protease-proteoglycan complexes: Tryptase resides in a complex distinct from chymase and carboxypeptidase. *J Immunol.* 1992;148:2475-82.
14. Alter SC, Kramps JA, Janoff A, Schwartz LB. Interactions of human mast cell tryptase with biological protease inhibitors. *Arch Biochem Biophys.* 1990;276:26-31.
15. Alter SC, Schwartz LB. Effect of histamine and divalent cations on the activity and stability of tryptase from human mast cells. *Biochim Biophys Acta.* 1989;991:426-30.
16. Stürzebecher J, Prasa D, Sommerhoff CP. Inhibition of human mast cell tryptase by benzamidine derivatives. *Biol Chem Hoppe Seyler.* 1992;373:1025-30.
17. Caughey GH, Raymond WW, Bacci E, Lombardy RJ, Tidwell RR. Bis(5-amidino-2-benzimidazolyl)methane and related amidines are potent, reversible inhibitors of mast cell tryptases. *J Pharmacol Exp Ther.* 1993;264:676-82.
18. Alter SC, Metcalfe DD, Bradford TR, Schwartz LB. Regulation of human mast cell tryptase. Effects of enzyme concentration, ionic strength and the structure and negative charge density of polysaccharides. *Biochem J.* 1987;248:821-7.
19. Schechter NM, Eng GY, McCaslin DR. Human skin tryptase: Kinetic characterization of its spontaneous inactivation. *Biochemistry.* 1993;32:2617-25.
20. Schwartz LB, Bradford TR, Littman BH, Wintroub BU. The fibrinogenolytic activity of purified tryptase from human lung mast cells. *J Immunol.* 1985;135:2762-7.

21. Lohi J, Harvima I, Keski-Oja J. Pericellular substrates of human mast cell tryptase: 72,000 Dalton gelatinase and fibronectin. *J Cell Biochem.* 1992;50:337-49.
22. Gruber BL, Marchese MJ, Suzuki K, et al. Synovial procollagenase activation by human mast cell tryptase dependence upon matrix metalloproteinase 3 activation. *J Clin Invest.* 1989;84:1657-62.
23. Malone DG, Wilder RL, Saavedra-Delgado AM, Metcalfe DD. Mast cell numbers in rheumatoid synovial tissues. Correlations with quantitative measures of lymphotic infiltration and modulation by anti-inflammatory therapy. *Arthritis Rheum.* 1987;30:130-7.
24. Hawkins RA, Claman HN, Clark RA, Steigerwald JC. Increased dermal mast cell populations in progressive systemic sclerosis: a link in chronic fibrosis. *Ann Intern Med.* 1985;102:182-6.
25. Schwartz LB. Tryptase from human mast cells: biochemistry, biology and clinical utility. *Monogr Allergy.* 1990;27:90-113.
26. Walls AF, Brain SD, Desai A, et al. Human mast cell tryptase attenuates the vasodilator activity of calcitonin gene-related peptide. *Biochem Pharmacol.* 1992;43:1243-8.
27. Tam EK, Caughey GH. Degradation of airway neuropeptides by human lung tryptase. *Am J Respir Cell Molec Biol.* 1990;3:27-32.
28. Hartmann T, Ruoss SJ, Raymond WW, Seuwen K, Caughey GH. Human tryptase as a potent, cell-specific mitogen: Role of signaling pathways in synergistic responses. *Am J Physiol Lung Cell Mol Physiol.* 1992;262:L528-34.
29. Miller JS, Westin EH, Schwartz LB. Cloning and characterization of complementary DNA for human tryptase. *J Clin Invest.* 1989;84:1188-95.
30. Miller JS, Moxley G, Schwartz LB. Cloning and characterization of a second complementary DNA for human tryptase. *J Clin Invest.* 1990;86:864-70.
31. Vanderslice P, Ballinger SM, Tam EK, Goldstein SM, Craik CS, Caughey GH. Human mast cell tryptase: Multiple cDNAs and genes reveal a multigene serine protease family. *Proc Natl Acad Sci USA.* 1990;87:3811-5.
32. Huang R, Åbrink M, Gobl AE, et al. Expression of a mast cell tryptase in the human monocytic cell lines U-937 and Mono Mac 6. *Scand J Immunol.* 1993;38:359-67.
33. Blom T, Hellman L. Characterization of a tryptase mRNA expressed in the human basophil cell line KU812. *Scand J Immunol.* 1993;37:203-8.
34. Reynolds DS, Gurley DS, Austen KF, Serafin WE. Cloning of the cDNA and gene of mouse mast cell protease-6. Transcription by progenitor mast cells and mast cells of the connective tissue subclass. *J Biol Chem.* 1991;266:3847-53.
35. McNeil HP, Reynolds DS, Schiller V, et al. Isolation, characterization, and transcription of the gene encoding mouse mast cell protease 7. *Proc Natl Acad Sci USA.* 1992;89:11174-8.
36. Sarid J, Benfey PN, Leder P. The mast cell-specific expression of a protease gene, RMCP II, is regulated by an enhancer element that binds specifically to mast cell trans-acting factors. *J Biol Chem.* 1989;264:1022-6.

37. Johnson DA, Barton GJ. Mast cell tryptases: examination of unusual characteristics by multiple sequence alignment and molecular modeling. *Protein Sci.* 1992;1:370-7.
38. Schwartz LB, Lewis RA, Seldin D, Austen KF. Acid hydrolases and tryptase from secretory granules of dispersed human lung mast cells. *J Immunol.* 1981;126:1290-4.
39. Craig SC, Irani AA, Metcalfe DD, Schwartz LB. Ultrastructural localization of heparin to human mast cells of the MC_{TC} and MC_T types by labeling with antithrombin III-gold.. *Lab Invest.* 1993;69:552-61.
40. Craig SS, Schechter NM, Schwartz LB. Ultrastructural analysis of maturing human T and TC mast cells in situ. *Lab Invest.* 1989;60:147-57.
41. Irani AA, Nilsson G, Miettinen U, et al. Recombinant human stem cell factor stimulates differentiation of mast cells from dispersed human fetal liver cells. *Blood.* 1992;80:3009-21.
42. Mitsui H, Furitsu T, Dvorak AM, et al. Development of human mast cells from umbilical cord blood cells by recombinant human and murine C-kit ligand. *Proc Natl Acad Sci USA.* 1993;90:735-9.
43. Soter NA, Wasserman SI, Austen KF, McFadden ER, Jr. Release of mast-cell mediators and alterations in lung function in patients with cholinergic urticaria. *N Engl J Med.* 1980;302:604-8.
44. Furitsu T, Saito H, Dvorak A, et al. Development of human mast cells in vitro. *Proc Natl Acad Sci USA.* 1989;86:10039-43.
45. Irani AA, Craig SS, Nilsson G, Ishizaka T, Schwartz LB. Characterization of human mast cells developed *in vitro* from fetal liver cells cocultured with murine 3T3 fibroblasts. *Immunology.* 1992;77:136-43.
46. Wenzel S, Irani AM, Sanders JM, Bradford TR, Schwartz LB. Immunoassay of tryptase from human mast cells. *J Immunol Methods.* 1986;86:139-42.
47. Enander I, Matsson P, Nystrand J, et al. A new radioimmunoassay for human mast cell tryptase using monoclonal antibodies. *J Immunol Methods.* 1991;138:39-46.
48. Schwartz LB, Metcalfe DD, Miller JS, Earl H, Sullivan T. Tryptase levels as an indicator of mast-cell activation in systemic anaphylaxis and mastocytosis. *N Engl J Med.* 1987;316:1622-6.
49. Wenzel SE, Fowler AA, 3d., Schwartz LB. Activation of pulmonary mast cells by bronchoalveolar allergen challenge. In vivo release of histamine and tryptase in atopic subjects with and without asthma. *Am Rev Respir Dis.* 1988;137:1002-8.
50. Castells M, Schwartz LB. Tryptase levels in nasal-lavage fluid as an indicator of the immediate allergic response. *J Allergy Clin Immunol.* 1988;82:348-55.
51. Shalit M, Schwartz LB, Golzar N, et al. Release of histamine and tryptase in vivo after prolonged cutaneous challenge with allergen in humans. *J Immunol.* 1988;141:821-6.
52. Schwartz LB, Atkins PC, Bradford TR, Fleekop P, Shalit M, Zweiman B. Release of tryptase together with histamine during the immediate cutaneous response to allergen. *J Allergy Clin Immunol.* 1987;80:850-5.

53. Butrus SI, Ochsner KI, Abelson MB, Schwartz LB. The level of tryptase in human tears: an indicator of activation of conjunctival mast cells. *Ophthalmology.* 1990;97:1678-83.
54. Francis H, Awadzi K, Ottesen EA. The Mazzotti reaction following treatment of onchocerciasis with diethylcarbamazine: clinical severity as a function of infection intensity. *Am J Trop Med Hyg.* 1985;34:529-36.
55. Schwartz LB, Yunginger JW, Miller JS, Bokhari R, Dull D. The time course of appearance and disappearance of human mast cell tryptase in the circulation after anaphylaxis. *J Clin Invest.* 1989;83:1551-5.
56. Van der Linden PG, Hack CE, Poortman J, Vivié-Kipp YC, Struyvenberg A, Van der Zwan JK. Insect-sting challenge in 138 patients: Relation between clinical severity of anaphylaxis and mast cell activation. *J Allergy Clin Immunol.* 1992;90:110-8.
57. Yunginger JW, Nelson DR, Squillace DL, et al. Laboratory investigation of deaths due to anaphylaxis. *J Forensic Sci.* 1991;36:857-65.
58. Platt MS, Yunginger JW, Sekula-Perlman A, Irani AA, Schwartz LB. Involvement of mast cells in sudden infant death syndrome. *J Allergy Clin Immunol.* 1994;in press
59. Sedgwick JB, Calhoun WJ, Gleich GJ, et al. Immediate and late airway response of allergic rhinitis patients to segmental antigen challenge: Characterization of eosinophil and mast cell mediators. *Am Rev Respir Dis.* 1991;144:1274-81.
60. Bosso JV, Schwartz LB, Stevenson DD. Tryptase and histamine release during aspirin-induced respiratory reactions. *J Allergy Clin Immunol.* 1991;88:830-7.
61. Broide DH, Eisman S, Ramsdell JW, Ferguson P, Schwartz LB, Wasserman SI. Airway levels of mast cell-derived mediators in exercise-induced asthma. *Am Rev Respir Dis.* 1990;141:563-8.
62. Schwartz LB, Bradford TR, Rouse C, et al. Development of a new, more sensitive immunoassay for human tryptase: use in systemic anaphylaxis. *J Clin Immunol.* 1994;11361:25957-7756.

*Biological and Molecular Aspects of Mast Cell
and Basophil Differentiation and Function,*
edited by Y. Kitamura, S. Yamamoto, S.J. Galli, and
M.W. Greaves. Raven Press, Ltd., New York © 1995.

15
IMMUNE- AND MAST CELL-INTERACTIONS WITH NERVES

John Bienenstock

*Intestinal Disease Research Programme
Department of Pathology
and
Molecular Virology and Immunology Programme
McMaster University
Hamilton, Ontario
L8N 3Z5
Canada*

It has been known for centuries that the mind may influence the body. Indeed, it was Francis Bacon[1] who maybe first expressed these ideas some 400 years ago. Some of these ideas are now fashionable, and somewhat unfortunately have been often been used in a diffuse and unrigorous approach to human disease. We all realize that emotions can affect the vascularity and even edema of the nasal mucosa and yet it has taken careful study of human gastric function as a result of exteriorization of the gastric mucosa to begin to appreciate the extent to which emotion can affect physiological function. The careful work of Wolf and Wolff[2] on human gastric function in 1944 clearly identified these associations. Their further work was to study and carefully document changes in the nasal mucosa of human subjects during varying life experiences. In this study published as a book in 1950[3], they successfully documented similar physiological activities in both stomach and nose as responses to anger, anxiety and resentment. Furthermore, happiness, depression, disgust, accompanied by helplessness showed similar responses in both mucosal tissues. It is perhaps not so well known that under the influence of emotion, leukocytes were induced to emigrate from the vascular bed into the nasal secretions. We would now recognize this as an example of the effects of brain-nerve interactions on the expression of endothelial adhesion molecules, and fit this into the now extensive literature[4], which in summary has shown bidirectional communication between the nervous and immune systems. This communication has been examined physiologically in situ, ex vivo as well as in vitro and has involved examination of tissue cells as well as circulating blood cells such as lymphocytes and their expression of receptors for various neuropeptides.[5,6,7,8,9]

The main subject of this symposium focusses on mast cells. We have highlighted in a number of publications the interactions between mast cells and nerves, nevertheless, it may be necessary to put these interactions into a broader context such as may occur in vivo under normal conditions. The reader will readily be able to identify from this examination many areas of potential research in which disordered physiological states may affect the brain and/or the nervous system directly or indirectly through effects on various components of the immune system.

While our own studies have largely been involved with the mucosal immune system, we believe that the lessons to be learned from these interactions have very broad implications, and both could, and should be extended beyond them to other tissues.

MAST CELL/NERVE INTERACTIONS

While we focus in this section on mast cell/nerve interactions, we believe that anatomical associations between nerves and plasma cells, eosinophils and other cells in the rat jejunum have both morphologic and functional consequences.[10,11] Neuropeptides are known to directly affect a whole spectrum of immune functions such as natural killer cell activity[12] and immunoglobulin synthesis.[8]

The size of the enteric nervous system is enormous. There are said by Furness and Costa[13] to be approximately 10^{14} cell bodies in this "little brain" capable of synthesis of at least 30-40 different neurotransmitters. The size and ramification of the nervous system affords a considerable opportunity for cells of the immune system and nerves to interact. Significant numbers of mast cells have been shown to be non-randomly apposed to mast cells in the jejunum of healthy and nematode-infected rats.[11] The apposing nerves were shown to contain substance P (SP) and/or calcitonin gene-related peptide. Similar results have been found by others in the rat[15] and also in the human.[16] Co-culture experiments between murine superior cervical ganglia and rat basophil leukemic cells (RBL-2H3, analogous to mucosal mast cells) or peritoneal mast cells have shown both tropism of nerves toward these cells, contact formation, branching and very close associations between apparent nerve endings and mast cells.[17,18] The consequences of these interactions have also been explored electrophysiologically. Many other examples of mast cell and other immune cell apparent morphometric associations in a variety of tissues including the intestine have been amply and extensively documented.[9]

Functional Considerations

In animals sensitized to a variety of food or other antigens, intestinal tissues display an increase in epithelial short circuit current (indicative of net ion transport) when challenged with the relevant antigen in Ussing chambers.[19,20] This response appears to be mediated by mast cells and several of their products such as arachidonic acid metabolites, and mediators such as histamine and serotonin, after cross linkage of mast cell bound IgE. Some of this effect is nerve dependent as shown by its tetrodotoxin sensitivity,[21,19] implying that this is an axon reflex between mast cells, nerves and the mucosal epithelium. Further studies in the W/Wv[20] mouse strain, which lacks mast cells because of a deficiency of the c-kit receptor, have provided unequivocal data in this regard since electrical field stimulation of enteric nerves showed a major reduction in the response in the W/Wv animals. This electrical short circuit current response was normalized by reconstitution of mast cells with adoptive transfer of bone marrow precursor cells. Lastly, in guinea pigs sensitized to antigen, specific antigen challenge of the intestine showed the expected short circuit current change accompanied by the simultaneous release of acetylcholine.[22] Since this response could be reproduced by the application of histamine, this suggested that antigen activation of mast cells lead to changes in nerve function which affected mast cells. Extensive examination of the role of mast cells and their products especially histamine by Wood, Weinreich and Undem[23,24] have shown that mast cell derived histamine may have a dual effect on the pre-synaptic and post-synaptic membranes through H3 and H2 receptors respectively.

Others have shown that neurostimulation can have variable effects, both on the mast cells of the intestine[25] and also those of the dura mater and tongue.[26] The effects of ipsilateral stimulation of the trigeminal ganglion on mast cells were dependent upon the intensity and extent in time of stimulation. Degranulation only occurred after prolonged stimulation and this same effect has also been seen only after prolonged stimulation of the saphenous nerve using antidromic stimulation.[27]

We have recently examined the effects of various concentrations of Substance P on rat peritoneal mast cells.[28] We have found using patch clamping of the mast cells that Substance P causes outwardly rectified chloride dependent oscillatory currents at physiological concentrations such as 50 nM.[29] No degranulation of mast cells was seen under these conditions. Similar reactive responses to the application of Substance P occurred at 5 pM and again no degranulation occurred. Repetition of these sub-threshold responses even at 5 pM caused degranulation to occur but only after a considerable delay of more than 25 minutes. These delays were characteristic of the concentrations of Substance P used: the lower the concentration, the greater the refractory period before degranulation. Thus, we believe that, when taken together with other studies which have shown that Substance P may prime mast cells for degranulation to subsequent sub-threshold anti-IgE, these results suggest that one of the functional correlates of the association of mast cells with nerves may well be the priming of these cells to respond to lower levels of normal stimuli. This is especially interesting since it requires supraphysiological concentrations of Substance P to degranulate mast cells in vitro. Similar priming effects of various growth factors and cytokines on basophils and mast cells have been shown.[30,31] The effect of repetition of the priming agents has not been previously reported. It is important to recognize that it is possible to cause mast cell synthesis and release of mediators and/or potential mediators without associated degranulation. For example, LPS can cause rat peritoneal mast cells to release IL-6 without concomitant histamine secretion.[32] Thus, neurotransmitters may have potent effects on cells without overt degranulation. Indeed overt degranulation may be the end of the line response for mast cells and our past efforts to concentrate attention on these events may have obscured even more physiologically important events at an earlier stage of activation of these cells. Mast cells sensitized with either IgG or IgE can respond to cross linking of these surface receptors and this can be influenced by a variety of local inflammatory effects including a variety of cytokines, growth factors and neuropeptides. Since mast cells are known to contain a number of factors which have been shown to have the capacity to communicate with nerves, it would not be surprising, and might be indeed entirely plausible, to begin to consider that such immune cells in contact with the peripheral nervous system can act as sensory receptors for antigen. This would be a particular advantage in mucosal tissues and skin, since such mechanisms would allow portions of the brain to appreciate not only the amount and extent of environmental antigens impinging on these tissues, but also the extent of reaction and inflammation consequent to these. Thus, taken in concert with the possibility that the nervous system may up or down regulate these cells, it should be possible to harness these systems and regulate them to promote the appropriate local responses.

It is not always true that stimulation of mast cells, either directly through nerves or via the secretion of neurotransmitters will cause (positive) degranulatory changes as described in mast cells. Inhibitory nerves to mast cell degranulation, especially of the NANC variety, have been described.[33] Upon stimulation of vagus nerves in cats sensitized to ascaris who inhaled ascaris antigen, the expected changes in histamine and pulmonary resistance did not occur if the vagus nerves were stimulated at the same time as administration of antigen.

Immune Cells Influence Neuronal Growth - Role of Nerve Growth Factor and Cytokines

Clearly, mast cells are involved in the triggering or maintenance of a variety of inflammatory and immune responses. The administration of Il-1 leads to increase of noradrenaline metabolism in the rat hypothalamus.[10] Enhanced neuronal survival in culture occurs and both the message and product for nerve growth factor (NGF) and its receptor are upregulated as a response to Il-1.

Recent demonstrations[34] that rat peritoneal mast cells can express the mRNA for leukemia inhibitory factor (LIF) and that mast cell lines released bioactive LIF, taken together with the known cholinergic differentiation factor activity of LIF, suggest that mast cells may influence the phenotype and growth of developing neurites in the course of inflammation. NGF is a pleiotropic polypeptide that influences many immune functions. Administration of NGF to neonatal mice causes increases in mast cell numbers[35,36] and the increase in connective tissue mast cells in this model have been shown to be due to mast cell degranulation.[37] The increases in mucosal mast cells in the same model are probably due to an indirect effect of NGF possibly through an effect on T-lymphocytes.[38] Since there are known changes which occur in neuronal remodelling in the course of inflammation in the intestine,[39] we set up a co-culture model of rat and mouse superior cervical ganglia with thymus, spleen and mesenteric lymph node (MLN) explants from normal and nippostrongylus infected animals.[40] The system examined the number of neurites which grew out towards the explants and in this way it was shown that both thymocytes and splenocytes promoted NGF dependent neurite growth which did not occur in the presence of normal MLN. MLN from infected rats showed a time course dependent enhanced neurite growth only 8% of which was inhibitable by anti-NGF antibody. Followup studies in the mouse reproduced these findings and showed in addition with specific antibodies to a variety of cytokines that the thymocyte and splenocyte dependent enhancement of neurite growth were mediated both by IL-1 and NGF. It is yet to be shown which cells are making either of these factors in this system. Interestingly, the cytokines responsible for the NGF independent growth in MLN from infected animals were IL-1, IL-3, GM-CSF and IL-6 all of which seem to contribute to this neurite growth.

Thus, there is clear evidence for bidirectional communication between nerves and cells of the immune response including mast cells. Nerve growth factors of a variety of sorts including NGF, LIF, IL-1, IL-6, GM-CSF and IL-3 play a variable role in remodelling of the nervous system associated with inflammation. Since many of these factors also interact with mast cells in a potent and meaningful way, it is clear that examination of hypersensitivity reactions in tissues must be done with these multiple and complex potential interactions in mind in order to establish the importance and significance of these kinds of observations.

CONCLUSIONS

Mast cells have been shown to be apposed to nerves throughout the body. Evidence for a bidirectional communication pathway exists and comes from experiments in vivo and ex vivo and also from co-culture experiments involving mast cells and neurites from superior cervical ganglia. Patch clamping of mast cells has shown that very low concentrations of the neurotransmitter substance P (5pM) will cause electrical oscillatory currents in the mast cell without degranulation. Repetition of the stimulus causes degranulation which normally only occurs with micro molar concentrations. These experiments suggest that one functional reason for the apposition of nerves and mast cells may be to enhance the sensitivity of the mast cell-nerve communication by lowering the threshold of response.

Since nerves appear to be constantly remodelled in situations where inflammation occurs, we examined the role of various lymphoid tissues in neurite outgrowth and survival. Explants of spleen, thymus and mesenteric lymph nodes caused neurite outgrowth from superior cervical ganglia via the formation of nerve growth factor and IL-1. Other cytokines may play a role in addition in the course of inflammation, such as IL-3, IL-6 and GM-CSF. Since nerve growth factor is a potent stimulator of mast cell growth, these complex interactions may regulate the capacity of immune cells to communicate with each other.

ACKNOWLEDGEMENT:

The author wishes to express his gratitude for the continuing financial support from the Medical Research Council of Canada and Astra Pharma Canada.

REFERENCES

1. Bacon F. In: Of the Proficiencie and Advancement of Learning, Divine and Humane. 1605.
2. Wolf S, Wolff HG. Human gastric function. Oxford Med Publ 1944.
3. Holmes TH, Goodell H, Wolf S, Wolff HG. The nose. An experimental study of reactions within the nose in human subjects during varying life experiences. C.C. Thomas. Springfield 1950;1-153.
4. Bienenstock J, MacQueen G, Sestini P, Marshall JS, Stead RH, Perdue MH. Mast cell/nerve interactions in vitro and in vivo. *Am Rev Respir Dis* 1991;143:S55-S58.
5. Blalock JE. A molecular basis for bidirectional communication between the immune and the neuroendocrine systems. *Physiol Rev* 1989;69:1-32.
6. McKay DM, Bienenstock J. Mast cell-nerve interactions in the gastrointestinal tract. *Immunol Today* 1994;15:533-538.
7. Payan DG. Neuropeptides and inflammation: the role of substance P. *Annu Rev Med* 1989;40:341-352.
8. Stanisz AM, Befus D, Bienenstock J. Differential effects of vasoactive intestinal peptide, substance P, and somatostatin on immunoglobulin synthesis and proliferation by lymphocytes from Peyer's patches, mesenteric lymph nodes, and spleen. *J Immunol* 1986;136:152-156.
9. Stead RH, Bienenstock J. Cellular interactions between the immune and peripheral nervous systems. A normal role for mast cells? In: *Cell to Cell Interaction*. MM Burger, Sordat B, Zinkernagel RM, Karger AG, Basel 1990;170-187.
10. Brenneman DE, Schultzberg M, Bartfai T, Gozes I. Cytokine regulation of neuronal survival. *J Neurochem* 1992;2:454-460.
11. Stead RH, Tomioka G, Quinonez G, Simon GT, Felten SY, Bienenstock J. Intestinal mucosal mast cells in normal and nematode-infected rat intestines are in intimate contact with peptidergic nerves. *Proc Natl Acad Sci,* USA 1987;84:2975-2979.
12. Croitoru K, Ernst PB, Bienenstock J, Padol I, Stanisz AM. Selective modulation of the natural killer activity of murine intestinal intraepithelial leukocytes by the neuropeptide substance P. *Immunol* 1990;71:196-201.
13. Furness JB, Costa M. Types of nerves in the enteric nervous system. *Neurosci* 1980;5:1-20.

14. Bienenstock J. Cellular communication networks: Implications for our understanding of gastrointestinal physiology. In: Stead RH, Perdue MH, Cooke H, Powell DW, Barrett KE. *Neuro-Immuno-Physiology of the Gastrointestinal Mucosa: Implications for Inflammatory Diseases.* NY Acad Sci, New York 1992;664:1-9.
15. Arizono N, Matsuda S, Hattori T, Kojima Y, Maeda T, Galli SJ. Anatomical variation in mast cell nerve associations in the rat small intestine, heart, lung, and skin: similarities of distances between neural processes and mast cells, eosinophils, or plasma cells in the jejunal lamina propria. *Lab Invest* 1990;62:626-634.
16. Stead RH, Dixon MF, Bramwell NH, Riddell RH, Bienenstock J. Mast cells are closely apposed to nerves in the human gastrointestinal mucosa. *Gastroenterol* 1989;97:575-585.
17. Blennerhassett MG, Tomioka M, Bienenstock J. Formation of contacts between mast cells and sympathetic neurons in vitro. *Cell Tissue Res* 1991;265:121-128.
18. Blennerhassett MG, Bienenstock J. Apparent innervation of rat basophilic leukaemia (RBL-2H3) cells by sympathetic neurons in vitro. *Neurosci Lett* 1990;120:50-54.
19. Castro GA, Harari Y, Russell D. Mediators of anaphylaxis-induced ion transport changes in small intestine. *Amer J Physiol* 1987;253:G540-G548.
20. Perdue MH, Masson S, Wershil BK, Galli SJ. Role of mast cells in ion transport abnormalities associated with intestinal anaphylaxis. Correction of the diminished secretory response in genetically mast cell-deficient W/Wv mice by bone marrow transplantation. *J Clin Invest* 1991;2:687-693.
21. Baird AW, Cuthbert AW. Neuronal involvement in type 1 hypersensitivity reactions in gut epithelia. *Br J Pharmacol* 1987;92:647-655.
22. Javed NH, Cooke HJ. Acetylcholine release from colonic submucous neurons associated with chloride secretion in the guinea pig. *Am J Physiol* 1992;262:G131-G136.
23. Weinreich D, Undem BJ. Immunological regulation of synaptic transmission in isolated guinea pig autonomous ganglia. *J Clin Invest* 1987;79:1529-1532.

24. Wood JD. Histamine signals in enteric neuroimmune interactions. *Ann NY Acad Sci* 1992;664:275-283.
25. Bani-Sacchi T, Barattini M, Bianchi S, Blandina P, Brunelleschi S, Fantozzi R, Mannaioni PF, Masini E. The release of histamine by parasympathetic stimulation in guinea pig auricle and rat ileum. *J Physiol* 1986;371:29-43.
26. Dimitriadou V, Buzzi MG, Moskowitz MA, Theoharides TC. Trigeminal sensory fiber stimulation induces morphological changes reflecting secretion in rat dura mater cells. *Neurosci* 1991;44:97-112.
27. Kowalski ML, Kaliner MA. Neurogenic inflammation, vascular permeability, and mast cells. *J Immunol* 1988;140:3905-3911.
28. Shanahan F, Denburg JA, Fox J, Bienenstock J, Befus D. Mast cell heterogeneity: Effects of neuroenteric peptides on histamine release. *J Immunol* 1985;135:1331-1337.
29. Janiszewski J, Bienenstock J, Blennerhassett MG. Picomolar doses of SP trigger electrical responses in mast cells without degranulation. *Am J Physiol*. 1994;C138-C145.
30. Bischoff SC, Brunner T, deWeck AL, Dahinden CA. Interleukin 5 modifies histamine release and leukotriene generation by human basophils in response to diverse agonists. *J Exp Med* 1990;172:1577-1582.
31. Bischoff SC, deWeck AL, Dahinden CA. Interleukin 3 and granulocyte/macrophage-colony-stimulating factor render human basophils responsive to low concentrations of complement component C3a. *Proc Natl Acad Sci, USA*. 1990;87:6813-6817.
32. Leal-Berumen I, Conlon P, Marshall JS. IL-6 production by rat peritoneal mast cells is not necessarily preceded by histamine release and can be induced by bacterial lipopolysaccharide. *J Immunol*. In Press.
33. Miura M, Inoue H, Ichinose M, Kimura K, Katsumata U, Takishima T. Effect of nonadrenergic noncholinergic inhibitory nerve stimulation on the allergic reaction in cat airways. *Am Rev Respir Dis* 1990;141:29-32.
34. Marshall JS, Gauldie J, Nielsen L, Bienenstock J. Leukemia inhibitory factor production by rat mast cells. *Eur J Immunol* 1993;23:2116-2120.
35. Aloe L, Levi-Montalcini R. Mast cell increase in tissues of neonatal rats injected with the nerve growth factor. *Brain Research* 1977;133:358-366.

36. Levi-Montalcini R, Calissano P. Nerve growth factor as a paradigm of other polypeptide growth factors. *Trends in Neurosci* 1986;9:473-477.
37. Marshall JS, Stead RH, McSharry C, Nielsen L, Bienenstock J. The role of mast cell degranulation products in mast cell hyperplasia. I. Mechanism of action of nerve growth factor. *J Immunol* 1990;144:1886-1892.
38. Matsuda H, Coughlin MD, Bienenstock J, Denburg JA. Nerve growth factor promotes human hemopoietic colony growth and differentiation. *Proc Natl Acad Sci*, USA 1988;85:6508-6512.
39. Stead RH, Kosecka-Janiszewska U, Oestreicher AB, Dixon MF, Bienenstock J. Remodelling of B-50(GAP-43)- and NSE-immunoreactive mucosal nerves in the intestines of rats infected with Nippostronglyus brasiliensis. *J Neurosci* 1991;11:3809-3821.
40. Kannan Y, Stead RH, Goldsmith CH, Bienenstock J. Lymphoid tissues induce NGF-dependent and NGF-independent neurite outgrowth from rat superior cervical ganglia explants in culture. *J Neurosci Res* 1994;37:374-383.

16

Skin mast cell activation by autoantibodies in urticaria and therapeutic implications

Michihiro Hide, David M Francis, Robert M Barr and
Malcolm W. Greaves

*St John's Institute of Dermatology, UMDS,
St Thomas's Hospital, London, SE1 7EH, UK*

Human mast cells and basophils play a pivotal role in immediate type hypersensitivity, and are involved not only in allergic disorders but also in the pathogenesis of a variety of inflammatory disorders (reviewed in ref 1) such as psoriasis and inflammatory bowel disease[1]. Urticaria is a good example of mast cell mediated diseases, especially for type I allergy. It is characterised by recurrent wheal and flare reactions with itching, which can be mimicked by intradermal injection of antigens or the mast cell-derived mediator, histamine[2]. That histamine plays an important role in urticaria is supported by the clinical effectiveness of histamine receptor antagonists[3] and the finding of elevated blister fluid histamine levels in affected and clinically normal skin in chronic urticaria[4]. The mechanisms of mast cell activation, however, have remained obscure in the majority of cases of chronic urticaria. Antigens, which bind and cross-link specific IgE on the high affinity IgE receptors (FcεRI) of mast cells and basophils, are identified as a cause in only limited cases of this disease[5]. Patients are generally convinced that they are reacting to a dietary constituent, or that their symptoms are due to "stress". Emotional stress may aggravate urticaria but the role of food factors is controversial. It has been demonstrated in several studies[6,7,8] that enhanced whealing reactions occur in skin of patients with urticaria in response to injection of a wide rage of vasoactive agents, including histamine and vasoactive intestinal peptide. These studies point to a non-specific increase in the sensitivity of the cutaneous vasculature to vasopermeability-enhancing agents, and suggest the involvement of vasoactive mediators in urticaria. However, these studies tell us little about the pathogenic mechanism, especially about the factors which initially trigger mast cell activation.

The possibility of circulating histamine releasing factors being involved in the pathogenesis of chronic urticaria was first suggested by the finding of basophilic leucopenia in patients by Rorsman (1962)[9], and reduced sensitivity of peripheral basophils to anti-IgE as compared with healthy controls by Greaves et al. (1974)[10] and Kern and Lichtenstein (1976)[11]. More recently, we have identified the presence of histamine releasing autoantibodies, either anti-IgE and/or anti-FcεRI, in patients with chronic urticaria.

SKIN REACTION UPON INTRADERMAL INJECTION OF AUTOLOGOUS SERUM

In 1986, Grattan et al. demonstrated that intradermal injection of autologous serum induced a wheal and flare reaction in patients with chronic urticaria[12]. Electron microscopic study showed mast cell degranulation at the site of injection[13]. We have found positive skin reactions to intradermal injection of autologous serum in 95 of 161 (59%) patients with severe chronic urticaria. The serum skin test responses tended to follow clinical activity of the urticaria, being substantially reduced or abolished when patients were in remission[12]. Autologous skin tests with serum from healthy control subjects were negative (Table 1).

HISTAMINE RELEASING FACTOR IN SERA OF PATIENTS WITH CHRONIC URTICARIA

In order to characterise the serum factor causing positive skin test and mast cell degranulation *in vivo*, sera from such urticaria patients were tested for histamine releasing activity *in vitro* by incubation with basophil leukocytes obtained from healthy donors[14]. We have found that sera from 44 of 70 (63%) skin test positive patients released greater than 5% of the histamine content of basophils.

Pre-incubation of basophils with goat anti-IgE antibody or the patients serum, in the presence of EDTA, showed that they cross-desensitised the basophils to each other, suggesting that the serum factor caused induction of histamine release by an IgE- or IgE receptor-mediated mechanism.

TABLE 1. Autologous serum skin test in patients with chronic urticaria, symptomatic dermographism and apparent healthy controls.

	Total tested	Positive (wheal volume >9mm^3)
Chronic urticaria	161	95 (59%)
Symptomatic dermographism	6	0 (0%)
Apparent healthy control	18	0 (0%)

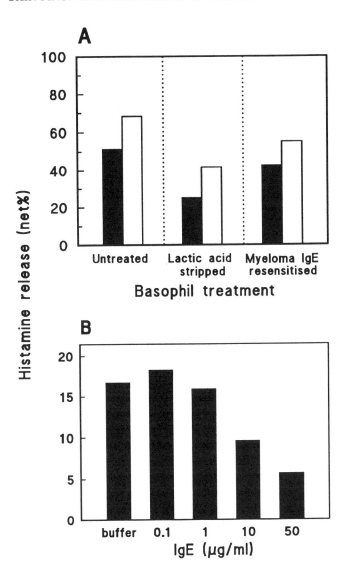

FIG. 1. A) Mean histamine release induced from endogenously IgE sensitised basophils of a healthy donor by sera of 5 patients with chronic urticaria. Effects of sequential removal of IgE by lactic acid stripping and resensitisation with myeloma IgE. Chronic urticaria sera, ■ ; goat anti-IgE, □ . B) Concentration dependent inhibition of urticaria patients' sera-induced histamine release from basophils by preincubation of the sera with myeloma IgE for 30 min at 37°C. Data are mean values for 3 sera.

Ultrafiltration of patients' sera indicated that the basophil histamine releasing activity was confined to the >100kD fraction, and affinity chromatography of serum on immobilised Protein G demonstrated that the activity was mainly attributable to IgG, suggesting an autoantibody mediated mechanism for the stimulation of histamine release.

Furthermore, we have demonstrated histamine release from dermal mast cells in skin slices incubated *in vitro* with sera from skin test positive patients and their purified IgG fraction[15].

HISTAMINE RELEASING AUTOANTIBODIES

Autologous IgG against IgE

Initially, cell bound IgE was identified as a target for histamine releasing autoantibodies in the sera of some patients with chronic urticaria[14]. Histamine release from normal basophils by the patients' sera and by goat anti-IgE polyclonal antibody was reduced to a similar extent following removal of cell bound IgE by lactic acid treatment. Almost complete restoration of histamine release was observed when the lactic acid treated cells were subsequently sensitized with myeloma IgE (Fig.1A). Furthermore, preincubation of the patients' sera with myeloma IgE (1-50µg/ml) also reduced the histamine releasing activity in a dose dependent manner (Fig.1B).

The biological and pathological relevance of the anti-IgE autoantibodies in chronic urticaria remains to be defined. Anti-IgE autoantibodies have been detected by immunoassay in a number of allergic disorders, and even in non-allergic healthy individuals[16]. However, in many instances, they fail to release histamine and their functional role is not known. Anti-IgE autoantibodies may be neutralised by plasma IgE or inhibit IgE binding to its high affinity receptor, FcεRI. Histamine release could presumably occur if the balance of free and IgE-bound anti-IgE antibodies is optimal, and if the epitope for anti-IgE antibodies is not concealed when IgE binds to FcεRI.

Autologous IgG against FcεRI

Studies of sera from some patients with chronic urticaria revealed histamine release from basophils of a healthy donor with a very low serum IgE. The same basophils failed to release histamine in response to a polyclonal goat anti-human IgE capable of cross-linking IgE. This suggests stimulation of histamine release by a mechanism independent of cell-bound IgE. Sera of four such patients were studied in detail[17]. Passive sensitisation of the low IgE basophils with myeloma

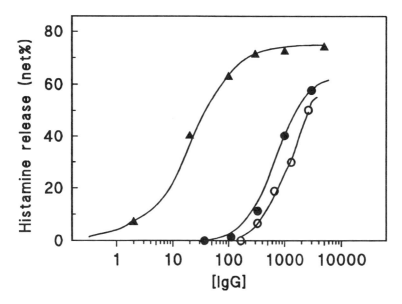

FIG. 2. Histamine release from non-IgE sensitised basophils of a low IgE donor by serum; ●, and purified IgG; ○, (μg/ml). Results are for a representative patient with chronic urticaria. Histamine release by 6F7 (ng/ml), ▲, an IgE competitive anti-FcεRIα monoclonal antibody[18], is shown for comparison. Goat anti-IgE released <5% histamine at an optimal dilution of 1/1000 (see Fig.3).

IgE, and its subsequent removal by lactic acid treatment, demonstrated that the patients sera induced greater histamine release when the cells were free of IgE. This was confirmed with endogenously IgE sensitized basophils from another donor, who had a high serum IgE concentration, by treating with lactic acid first and then re-sensitising with IgE.

Histamine releasing activities in the patients' sera were recovered in the IgG fraction on affinity chromatography on immobilised Protein G (Fig.2). Preincubation of the sera with a recombinantly produced fragment of the extracellular domain of the FcεRI α-subunit (sFcεRIα) effectively neutralised the histamine release induced from non-sensitized basophils (Fig.3). Serum histamine releasing activity in patients, for whom the clinical course of their urticaria could be followed, showed correlation with the severity of clinical symptoms. We have since confirmed the induction of histamine release from mast cells in human skin slices by the IgG fraction of the patients sera, and its inhibition by pre-incubation with sFcεRIα[15]. These results demonstrate the presence of functional histamine releasing autoantibodies against FcεRIα in chronic urticaria, and also imply a potential for the use of sFcεRIα in therapy.

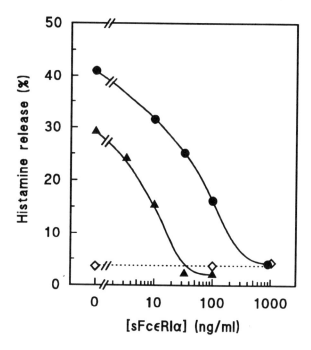

FIG. 3. Concentration dependent inhibition by sFcεRIα of histamine release from non-IgE sensitised basophils of a healthy donor by serum of a representative patient (same as for Fig. 2) with chronic urticaria, ● ; mouse anti-FcεRIα monoclonal antibody, 6F7, ▲ ; and goat anti-IgE antibody, ◇ . Serum, 6F7 and goat anti-IgE antibody were preincubated with sFcεRIα for 30 min at 37°C.

HETEROGENIETY AND DISTRIBUTION OF AUTOANTIBODIES IN CHRONIC URTICARIA

Studies of the distribution of anti-IgE and anti-FcεRIα autoantibodies in a large population of chronic urticaria patients are in progress. The relation between histamine releasing activity on IgE sensitized basophils (anti-IgE and/or non-IgE competitive anti-FcεRIα like activity) and that on non-IgE sensitised basophils (anti-FcεRI like activity) amongst a group of 42 patients is shown in Fig.4. A small proportion of sera released histamine from IgE sensitized, but not (<5% release) from non-IgE sensitized cells, therefore mimicking authentic anti-IgE antibodies. Another small group of sera released histamine from non-IgE sensitized basophils, but not from sensitised cells, thereby resembling the functional activity of 6F7, a mouse anti-human FcεRI, IgE competitive monoclonal antibody. A majority, approximately two thirds, of the sera released histamine from both types of

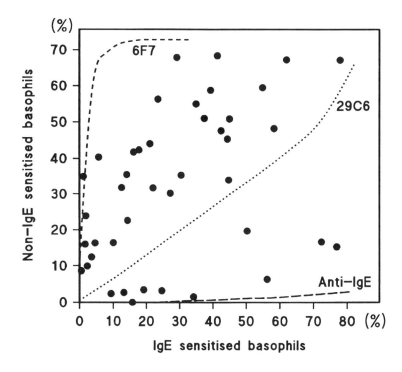

FIG. 4. Histamine release from IgE sensitized and non-IgE sensitized basophils induced by sera of chronic urticaria patients. 6F7 is an IgE competitive monoclonal antibody and 29C6 an IgE non-competitive monoclonal antibody against FcεRIα[18].

basophils, suggesting the presence of non- or partially IgE competitive type anti-FcεRI autoantibodies or coexistence of anti-FcεRI and anti-IgE autoantibodies. Preincubation of the sera of 10 patients in this group with sFcεRIα neutralised a major part of, if not all, histamine releasing activity[15] indicating that autoantibodies directed against FcεRIα represent the major mechanism in the pathogenesis of whealing in chronic urticaria.

APPLICATION OF IMMUNOMODULATING THERAPY FOR SEVERE UNREMITTING URTICARIA AS AN AUTOIMMUNE DISEASE

Our results raise the possibility that at least in some patients, chronic "idiopathic" urticaria may have an autoimmune aetiology. Accordingly, several therapeutic studies have been undertaken for severely affected cases of urticaria for which other conventional treatments were not effective.

Removal of circulating autoantibodies by plasmapheresis induced temporary remission or improvement in 6 of 8 patients with chronic urticaria, who had positive autologous serum skin tests and *in vitro* histamine releasing activity[19]. Blood cellular histamine, which is subject to depletion by circulating autoantibodies, showed an inverse correlation with the *in vitro* serum histamine releasing activity and clinical symptoms in one subject followed for 9 months.

Six of 8 patients treated by infusion of high-dose immunoglobulin[20], which is thought to suppress autoantibody production, demonstrated a clinical improvement and a reduction in autologous intradermal serum reactions[21]. One patient studied in detail showed a decrease in IgG mediated histamine release *in vitro*, correlating with disease activity[22].

We have also studied treatment with low dose cyclosporin A (2.5-3.5mg/kg/day). Although its precise mechanism of action in urticaria remained to be defined, 9 of 12 patients with severe chronic urticaria resolved or improved during treatment for 4 weeks, with 7 showing a sustained improvement for at least a month after stopping cyclosporin therapy[23].

These treatments are, however, non-selective. The development of more specific therapy for the histamine releasing autoantibody against anti-FcεRIα appears to be feasible, since apparent concentrations of the autoantibodies in sera are low (<1μg/ml) and they are neutralised by relatively low concentrations of sFcεRIα[17]. That *in vivo* administration of sFcεRIα[24] or an FcεRIα-human IgG chimeric molecule[25] prevents PCA reactions suggests that development of therapeutic systems for selective removal or neutralisation of histamine releasing anti-FcεRIα autoantibodies may be effective. Antigen specific immunoadsorbtion of anti-dsDNA autoantibodies in systemic lupus erythematosus (SLE) has already been tried with success[26]. Precise mapping of the epitope for anti-FcεRIα autoantibodies and structure-based design by computerised molecular modelling of truncated peptide sequences may allow us to develop new drugs which specifically neutralise the autoantibodies.

Although mast cells and their mediators are thought to be involved in the pathogenesis of a variety of diseases, the mechanism initiating mast cell activation remains unclear in the majority of such diseases. Activation of FcεRIα by autoantibodies may occur in other diseases that can have urticaria-like lesions, such as SLE.

It is possible that abnormal activation or inactivation of various receptors by cross-linkage may be induced by autoantibodies which are not detectable by *in vitro* immunoreactivity[27]. Investigation of receptor cross-linking autoantibodies may reveal the pathogenesis and new therapeutic approaches in a number of diseases which have been designated as "idiopathic".

ACKNOWLEDGMENTS

We thank the Eleanor Naylor Dana Charitable Trust, the Kieckhefer Foundation and the Dunhill Trust for financial support.

REFERENCES

1. Eds. Kaliner MA, Metcalfe DD. The mast cell in health and disease. New York: Marcel Dekker; 1993.
2. Sheldon JM, Mathews KP, Lovell RG. The vexing urticaria problem: present concepts of etiology and management. *J Allergy* 1954;25:525-539.
3. Estelle F, Simmons FER. Recent advances in H_1 receptor antagonist treatments. *J Allergy Clin Immunol* 1990;86:995-9.
4. Kaplan AP, Horakova Z, Katz SI. Assessment of tissue fluid histamine levels in patients with urticaria. *J Allergy Clin Immunol* 1978;61:350-4.
5. Champion RH, Roberts SO, Carpenter RG, Roger JH. Urticaria and angio-oedema - A review of 554 patients. *Br J Dermatol* 1969;81:588-97.
6. Krause LB, Shuster S. Enhanced weal and flare response to histamine in chronic idiopathic urticaria. *Br J Clin Pharmac* 1985;20:486-488.
7. Juhlin L. Late phase reaction to platelet activating factor and kallikrein in urticaria. *Clin Exp Allergy* 1990;20,Suppl 4:9-10.
8. Smith CH, Atkinson B, Morris RW, Hayes N, Foreman JC, Lee TH. Cutaneous responses to vasoactive intestinal polypeptide in chronic idiopathic urticaria. *Lancet* 1992;339:91-3.
9. Rorsman H. Basophilic leucopenia in different forms of urticaria. *Acta Allergologica* 1962;17:168-84.
10. Greaves MW, Plummer VM, McLaughlan P, Stanworth DR. Serum and cell bound IgE in chronic urticaria. *Clin Allergy* 1974;4:265-71.
11. Kern F, Lichtenstein LM. Defective histamine release in chronic urticaria. *J Clin Invest* 1976;57:1369-77.
12. Grattan CEH, Wallington TB, Warin RP, Kennedy CTC, Bradfield JW. A serological mediator in chronic idiopathic urticaria - a clinical, immunological and histological evaluation. *Br J Dermatol* 1986; 114:583-90.
13. Grattan CEH, Boon AP, Eady RAJ, Winkelmann RK. The pathology of the autologous serum skin test response in chronic urticaria resembles IgE-mediated late phase reactions. *Int Arch Allergy Appl Immunol* 1990;93: 198-204.
14. Grattan CEH, Francis DM, Hide M, Greaves MW. Detection of circulating histamine releasing autoantibodies with functional properties of anti-IgE in chronic urticaria. *Clin Exp Allergy* 1992;21:695-704.

15. Francis DM, Niimi N, Hide M, Kermani F, Barr RM, Black AK, Greaves MW. Histamine release from human skin *in vitro* induced by sera containing IgG anti-FcεRIα autoantibodies from patients with chronic urticaria. *J Invest Dermatol* 1994; in press (abstr).
16. Twena DM, Marshall JS, Haeney MR, Bell EB. A survey of non-atopic and atopic children and adults for the presence of anti-IgE autoantibodies. *Clin Immunol Immunopathol* 1989;53:40-51.
17. Hide M, Francis DM, Grattan CEH, Hakimi J, Kochan JP, Greaves MW. Auto-antibodies against the high affinity IgE receptor as a cause of histamine release in chronic urticaria. *New Eng J Med* 1993;328:1599-604.
18. Riske F, Hakimi J, Mallamaci M, et al. High affinity human IgE receptor (FcεRI): analysis of functional domains of the α-subunit with monoclonal antibodies. *J Biol Chem* 1991;266:11245-51.
19. Grattan CEH, Francis DM, Slater NGP, Barlow RJ, Greaves MW. Plasmapheresis for severe unremitting, chronic urticaria. *Lancet* 1992;339:1078-80.
20. Gelfand EW. Intervention in autoimmune disorders: creation of a niche for intravenous gamma-globulin therapy. *Clin Immunol Immunopathol* 1989;53:S1-6
21. O'Donnell BF, Barlow RJ, Black AK, Greaves MW. Response of severe chronic urticaria to intravenous immunoglobulin (IVIG). *Br J Dermatol* 1994;131,Suppl 44:23 (abstr).
22. Kermani F, Niimi N, Francis DM, O'Donnell BF, Barr RM, Greaves MW. Anti-FcεRIα and other histamine releasing activity in a patient with chronic urticaria treated by intravenous immunoglobulin. *Allergologie* 1994;17:224 (abstr).
23. Barlow RJ, Black AK, Greaves MW. Treatment of severe chronic urticaria with cyclosporin A. *Eur J Dermatol* 1993;3:273-5.
24. Ra C, Kuromitsu S, Hirose T, Yasuda S, Furuichi K, Okumura K. Soluble human high affinity receptor for IgE abrogates the IgE-mediated allergic reaction. *Int Immunol* 1993;5:47-54.
25. Haak-Frendscho M, Ridgway J, Shields R, Robbins K, Gorman C, Jardieu P. Human IgE receptor α chain IgG chimera blocks passive cutaneous anaphylaxis reaction in vivo. *J Immunol* 1993;151:351-8.
26. Suzuki K, Hara M, Ishizuka T, et al. Continuous anti-dsDNA antibody apheresis in systemic lupus erythematosus. *Lancet* 1990;336:753-4.
27 Hide M., Greaves MW. Failure to tolerate high affinity IgE receptors: A novel mechanism in the pathogenesis of diseases that result from abnormal cell activation. *Ann Med* 1994;26:117-8.

17
Pharmacological Modulation of Human Mast Cell and Basophil Secretion *in Vitro* and *in Vivo*

Gianni Marone, Giuseppe Spadaro, and Arturo Genovese

*Division of Clinical Immunology and Allergy,
University of Naples Federico II, School of Medicine,
80131 Naples, Italy*

There is compelling evidence that human basophils and mast cells play a major role in the pathogenesis of hypersensitivity reactions through the elaboration of proinflammatory mediators (1) and various cytokines (2–5). This concept is now firmly based on a series of anatomical, immunological and pharmacological observations. The number of basophils in sputum from asthmatics rises just before an asthmatic attack (6) and after allergen–induced asthmatic responses (7). Mast cells and basophils have been identified in the lumen of human bronchi (8) and bronchoalveolar lavage (BAL) fluid (9–11). Mast cells have also been found in the mucosa and submucosa of bronchial biopsies from patients with bronchial asthma (12–14). Interestingly, basophils have been found post–mortem in the airway lumen, in the bronchial epithelium and in the submucosa of fatal asthma cases (15). In addition, annual changes in basophils have been correlated to airway responsiveness (16).

Mast cells and basophils are activated in various allergic skin disorders such as acute and chronic urticaria (17,18), atopic dermatitis (19), and contact dermatitis (20). Functional differences, namely altered basophil and/or mast cell releasability, have been documented in patients with atopic dermatitis (19), chronic urticaria (17,18), and bronchial asthma (21). Moreover, human basophils (5) and mast cells (2–4) can generate cytokines after immunological activation. In addition, activated basophils and mast cells express the ligand for CD40 (CD40L), which can provide the cell contact signal required for IgE synthesis by human B cells (22). These observations suggest that mast cells and basophils may therefore play a key role in allergic disorders, not only by producing inflammatory mediators, but also by directly regulating IgE synthesis independently of T cells.

Despite their obvious similarities, human peripheral blood basophils and tissue mast cells differ morphologically, ultrastructurally, immunologically, biochemically and pharmacologically (1,23,24). Basophils, like other granulocytes, differentiate and mature in the bone marrow, circulate in the blood, and are not normally found in connective tissues. In contrast, mature mast cells are ordinarily distributed throughout connective tissues (24) where they are often adjacent to blood and lymph vessels, near or within nerves, and beneath epithelial surfaces, such as those of the respiratory and gastrointestinal systems and the skin, which are exposed to environmental antigens. These and many other differences (1,24) suggest that basophils and mast cells may play different roles in various types and phases of allergic reactions involving the respiratory tract (25,26) and skin (26).

An important feature of basophils is their ability to migrate into certain inflammed tissues. For instance, they are recruited to the site of late–phase reactions and synthesize most of the mediators recovered in nasal lavage and skin blister models during late–phase reactions (25–29). Mast cells seem to be involved predominantly in immediate reactions, and basophils in the late phases (10,11,25–29).

Numerous studies have focused on the *in vitro* effects of drugs that modulate the release of mediators from these cell types isolated from different tissues. This approach has contributed to clarifying the biochemical mechanism by which these cells release mediators in response to immunological and non-immunological stimuli (30–32). Significant differences have been documented: a) between mast cells and basophils as regards the pharmacological agents that modulate the release of the preformed (histamine) and *de novo* synthesized (LTC_4 and PGD_2) mediators they produce (1); and b) between mast cells isolated from different anatomical sites (24). We are starting to evaluate the effects of drugs on the release of mediators from these cells *in vivo* (33,34). In this chapter we shall concentrate on agents that modulate the release of chemical mediators from human basophils and mast cells *in vitro* and *in vivo*, and that appear to hold promise for the treatment of inflammatory diseases.

PHOSPHODIESTERASE INHIBITORS AND ADENYLATE CYCLASE ACTIVATORS

It has long been appreciated that cyclic AMP (cAMP) is an important second messenger in a variety of cell systems and, in general, it is believed to inhibit the function of human inflammatory cells (35). Interesting studies have demonstrated that leukocytes from patients with atopic dermatitis have increased levels of cAMP phosphodiesterase (PDE) activity (36). This abnormality can be reversed by inhibition of cAMP PDE. Also the increased IL-4 production by atopic mononuclear cells *in vitro* correlated with enhanced PDE activity and was reversed by a selective type IV PDE inhibitor (36). These observations suggest PDE isozyme as a possible therapeutic target for the control of atopic skin disease.

At least five different classes of cAMP PDE isozymes have been proposed, each the product of distinct gene or a family of genes (37). Table I summarizes the currently accepted classification. These PDE isozymes differ in: 1) kinetic characteristics; 2) physical characteristics; 3) tissue distribution; 4) endogenous activators and inhibitors (e.g., Ca^{2+}/calmodulin, cGMP); and 5) subcellular distribution (37).

A wide variety of PDE inhibitors are available as pharmacological probes, falling into two main categories: 1) standard non-specific inhibitors of PDE activity such as theophylline and 3-isobutyl-1-methylxanthine; and 2) a newer category, selective for individual PDE isozymes. The latter comprises compounds such as rolipram, a selective inhibitor of the cAMP-specific PDE (PDE IV) and zaprinast, a selective inhibitor of the cGMP-specific PDE (PDE V) (37). Both cAMP PDE III and PDE IV have been identified in broken cell preparations of purified basophils.

TABLE I

Cyclic Nucleotide Phosphodiesterase (PDE) Isozymes

Family	Characteristics	Inhibitors
I	Ca^{2+}/Calmodulin dependent	Vinpocetine
II	cGMP-stimulated	None selective
III	cGMP-inhibited cAMP >> cGMP specific	Siguazodan Milrinone
IV	cAMP-specific cGMP-insensitive	Rolipram Tibenelast
V	cGMP-specific	Zaprinast Dipyridamole

Two studies in human basophils have examined the role of selective PDE inhibitors in the modulation of histamine and LTC_4 release. Columbo et al. described a dose–dependent inhibition of PAF–induced histamine release by rolipram, reaching more than 60% inhibition at a concentration above 10 nM, and no effect by the type III PDE inhibitor SK&F 95654 (38). Peachell et al. obtained similar findings with rolipram, which inhibited anti–IgE–induced histamine and LTC_4 release and increased cAMP levels in basophils (39). In contrast, mediator release was not inhibited by compound SK&F 95654 or zaprinast, the PDE V inhibitor. Rolipram enhances the inhibitory effect of forskolin, a direct activator of adenylate cyclase (40), on mediator release from basophils and cAMP accumulation (39). It thus appears that a rise in intracellular cAMP inhibits mediator release from basophils and that PDE IV is the isozyme mainly responsible for regulating cAMP content in these cells. Inhibitors of PDE IV also regulate mast cell (41), eosinophil (42) and neutrophil (43) function.

Thus, despite the present lack of in vivo studies, in vitro data suggest that selective inhibitors of PDE isozymes may prove useful in the management of allergic and inflammatory diseases in the near future. However, it should be remembered that two currently licensed PDE inhibitors, theophylline and pentoxifylline, generally considered nonselective in their activity on PDE enzymes, are clinically effective in different conditions. Pentoxifylline in particular, appears to have several important antiinflammatory and immunoregulatory effects (44,45).

Isoproterenol and other catecholamines inhibit IgE–mediated histamine release from basophils, presumably by binding to β–adrenergic receptors (35). We used fenoterol, a $β_2$–adrenergic agonist, to assess the presence of $β_2$ adrenergic receptors on human basophils and lung mast cells. Fenoterol rapidly and transiently inhibited antigen–induced histamine release from both cell types (46) and the dose–response inhibition curve was paralleled by a fenoterol–induced increase in cAMP levels in leukocytes. Unfortunately, the use of fenoterol has been reported to be associated with an increased risk of death or near death from asthma (47,48). Other $β_2$–agonists could, however, achieve some anti–inflammatory effect by virtue of their ability to inhibit mediator release from basophils and mast cells through the activation of adenylate cyclase. Work is starting to evaluate the effect of newer $β_2$ adrenergic agonists with interesting pharmacokinetic properties, such as salmeterol and formoterol, on the release of mediators from lung mast cells (49).

NON STEROIDAL ANTI–INFLAMMATORY DRUGS

Aspirin and other non steroidal anti–inflammatory drugs (NSAID) exacerbate asthma in about 20% of patients (50) and this has been related to their ability to potentiate the release of basophil and mast cell mediators (51). Pharmacological concentrations of NSAID such as indomethacin (INDO), meclofenamic acid (MCA), and acetylsalicylic acid (ASA) enhance antigen- and anti-IgE-induced histamine release (51,52), and potentiate the *de novo* synthesis of LTC_4 from basophils (53). This appears to be a general property of all NSAID tested (54), and their potency in enhancing IgE-mediated histamine release was significantly correlated with their capacity to inhibit the cyclo-oxygenase pathway (54).

Most of the currently available acidic NSAID are either carboxylic (e.g., ASA, INDO, naproxen, and ibuprofen) or enolic acid (e.g., phenylbutazone and piroxicam). Nimesulide (4-nitro-2-phenoxymethane sulfonanilide; NIM), is an NSAID chemically unrelated to other compounds of the same class, its functional acid group being sulfonanilide (55). We tested the *in vitro* effects of NIM on basophils and mast cells isolated from lung parenchyma and skin tissues. Pharmacological concentrations (10^{-6} to $10^{-4} M$) of NIM and of its active metabolite 4-hydroxy-nimesulide (OH-NIM), concentration-dependently inhibited the release of histamine and LTC_4 from basophils activated by antigen and anti-IgE (56). NIM significantly inhibited the release of histamine from basophils, whereas INDO, MCA and ASA enhanced IgE-mediated histamine release from basophils (51). NIM also inhibited histamine, LTC_4 and PGD_2 release from lung parenchymal and skin mast cells (57).

This drug and its metabolite, differently from other NSAID, inhibit the release of preformed and *de novo* synthesized proinflammatory mediators from human FcεRI⁺ cells challenged with IgE- and non-IgE-mediated stimuli. NIM and OH-NIM also potentiated, whereas ASA and INDO reversed, the inhibitory effect of adenylate cyclase agonists such as prostaglandin E_1 and forskolin (57). NIM appears to be the only NSAID with these properties, which might explain why it is well tolerated in patients with ASA idiosyncrasy (58). Additional studies are necessary to characterize its biochemical mechanism of action in the light of the recent identification of two isozymes of cyclo-oxygenase (type I and II) (59).

CORTICOSTEROIDS

Endogenous glucocorticoid hormones play an important role in the regulation of the immune system. At pharmacological dosages, they are one of the therapeutic mainstrays in most inflammatory and immune-mediated disorders. However, the precise mechanisms by which corticosteroids exert these beneficial effects remain elusive, owing in part to their multiple biological actions.

The recent demonstration of inflammation in the bronchial mucosa even in patients with mild asthma (60-62) is leading to the wider use of inhaled and systemic corticosteroids. However, the mechanisms underlying their therapeutic activity in bronchial asthma and inflammatory disorders are still not completely understood. Intravenously injected corticosteroids in man cause rapid basopenia, whereas skin tissue histamine remains unchanged (63). Only prolonged treatment (at least four weeks) with topical corticosteroids reduced skin mast cell number and histamine content and inhibited the allergen–induced wheal–and–flare response (64). Short-term *in vitro* incubation with different corticosteroids did not inhibit mediator release from basophils, but prolonged (12 to 24 h) incubation with prednisolone, dexamethasone or deflazacort concentration–dependently inhibited IgE–mediated histamine release from these cells (65–67).

Interestingly, up to 24 h incubation of mast cells isolated from lung parenchyma, intestine or skin with dexamethasone did not alter their ability to release histamine, PGD_2 or LTC_4 after challenge with anti-IgE (68). This is consistent with observations that corticosteroids do not block the early phase of antigen challenge reactions in bronchial or nasal provocation (69), and provide a clear example of the pharmacological heterogeneity of human $Fc\epsilon RI^+$.

The fact that human basophils (5) and mast cells (2–4) are a major source of cytokines raises the important question whether corticosteroids can modulate the synthesis of cytokines in immunologically activated $Fc\epsilon RI^+$ cells.

Another important issue related to the use of systemic and topical corticosteroids in allergic patients is their well–known side effects. The potent topical corticosteroids recently introduced, such as budesonide and fluticasone, which are apparently extremely effective in the treatment of bronchial asthma (70), might possess side effects that have not yet been fully appreciated.

PROTEIN KINASE AND TYROSINE KINASE INHIBITORS

Activation of human basophils and mast cells by cross–linking of surface IgE involves an increase in the hydrolysis of cell membrane phosphatidyl–inositols (phosphoinositide turnover) to yield two intracellular messengers, inositol trisphosphate ($InsP_3$) and diacylglycerol (DAG). $InsP_3$ stimulates the release of Ca^{2+} from intracellular stores and DAG activates protein kinase C (PKC) (71,72). PKC plays a major role in signal transduction by phosphorylating different proteins including ion channels, receptors, other kinases and cytoskeletal proteins. In resting cells at least 90% of PKC activity is in the cytosol (73).

In rat basophilic leukemic cells, PKC activating agents such as phorbol ester (TPA) induced the rapid translocation of more than 55% of cytosolic PKC to the membrane, which correlates with receptor–induced exocytosis (74). TPA and bryostatins can activate human basophils, but not mast cells, to release histamine, but not LTC_4 (75,76). Anti-IgE- and TPA-induced activation of basophils is accompanied by a rapid rise in membrane-associated PKC, correlated with the amount of histamine released (77). Appropriate concentrations of staurosporine, a PKC inhibitor, inhibit TPA-induced mediator release from basophils (77). TPA and bryostatins both inhibit anti–IgE–induced mediator release from skin and lung mast cells (76,78). This only apparently contrasts with the observation that TPA and bryostatins do not activate human mast cells and suggests that, although PKC activation alone is not sufficient to cause mediator release from mast cells, it might contribute to the release induced by cross-linking of surface IgE.

The identification of several isoforms of PKC differing in their subcellular localization and their requirements for activation (79) may help in identifying novel molecular targets for the development of therapeutic agents useful in the management of basophil- and mast cell-mediated inflammatory diseases. In addition, PKC activation plays different roles in the various biochemical pathways induced by different secretagogues. Therefore, the effects of staurosporine and other PKC inhibitors differ with different secretagogues (80).

Recent evidence suggests that tyrosine kinases are important in signal transduction mechanisms in many cell types. In rodent mast cells and mast cell lines IgE cross-linking leads to an increase in tyrosine phosphorylation (81), suggesting that a tyrosine kinase may be involved in either the activation of phospholipase C or the regulation of long-term responses.

Four different inhibitors of tyrosine kinases were tested on IgE-dependent histamine release from human lung mast cells and basophils (82). Genistein and tyrphostin 25 inhibited histamine secretion from basophils, but had much less effect in lung mast cells. Methyl 2,5 dihydroxycinnamate did not inhibit mediator release from basophils, but proved to be an effective inhibitor of IgE-mediated degranulation in lung mast cells. Finally, lavendustin A failed to inhibit histamine release from either basophils or mast cells.

In summary, inhibitors of tyrosine kinases prevent IgE-dependent histamine secretion from human lung mast cells and basophils, though these two FcϵRI$^+$ cell types have different inhibitory profiles, possibly reflecting differences in IgE-dependent signal transduction mechanisms.

IMMUNOPHILIN LIGANDS

Cyclosporin A (CsA) is a cyclic undecapeptide isolated from a fungal extract that inhibits T cell-mediated immune responses (83). Since its clinical introduction in the late 70s, CsA has been credited with the dramatic increase in survival rates of patients receiving kidney, heart, and liver transplants. In addition clinical trials are now yielding encouraging results in a growing list of immune disorders (32,84,85).

CsA has recently offered some insight into the mechanisms of signal transduction in immune cells. The first of a family of proteins with high affinity for CsA was identified by Handschumacher *et al.* in 1984. This protein, named cyclophilin, displayed high affinity for CsA and low affinity for inactive analogs such as CsH (86). CyP belongs to a family of intracellular proteins, the immunophilins, which is emerging as a new pathway of intracellular signaling in immune and inflammatory cells. Two major classes, the cyclophilins (CyP) and FK-binding proteins (FKBP), have been identified, purified, and cloned (87).

At least four separate human isoforms of CyP have been identified, and appear to be products of a multigene family (88). Two natural macrolides, FK-506 and rapamycin (RAP) bind with high affinity to FKBP, including FKBP-12 and FKBP-13 (89). Although their amino acid sequences are unrelated, both immunophilins possess peptidyl–prolyl *cis–trans* isomerase activity [90], which is inhibited by their ligands, CsA and FK-506 (91).

It is clear that CsA binding to CyP, and FK-506 to FKBP is essential in their immunosuppressive activity (34). Complexes between CsA and CyP, as well as FK-506-FKBP complexes, bind to the set of proteins now identified as calcineurin (92). Calcineurin (Cn) has a catalytic A subunit (CnA) and a regulatory B subunit (CnB) (92). The CnA subunit, with the catalytic activity, has a binding site for calmodulin (CaM) and for the CnB subunit. Cn appears to have all the biochemical requirements of the common target of immunophilin–drug complexes and is thus a potential component of the signaling pathways involved in the activation of immune cells (93) (Table II).

TABLE II

The Immunological Significance of Calcineurin (Cn)/Phosphatase 2B Activity

- Cn is a key rate–limiting enzyme in T–cell signal transduction and in basophil/mast cell mediator release

- IL-2 production by T cells activated through the TCR/CD3 complex is correlated with the level of phosphatase 2B activity

- Cells expressing low levels of Cn (e.g., T cells) are more sensitive to CsA/FK-506

- Overexpression of Cn overcomes the CsA/FK-506-mediated inhibition of NF-AT-dependent cytokine gene transcription

- Immunosuppressive activity of cyclosporin analogs correlates with inhibition of Cn phosphatase 2B activity

- Cn is involved in signalling events that lead to degranulation of basophils, mast cells, and cytotoxic T cells

- Cn activity plays a role in TCR/CD3-mediated induction of apoptosis in T cell hybridomas

Complexes of CsA or FK–506 and their respective intracellular binding proteins inhibit the CaM–dependent protein phosphatase 2B, which appears to be essential in the signal transduction pathway for lymphocytes (93), human basophils (30,31,94–96) and mast cells (94,96,97).

RAP exerts distinct immunosuppressive activities presumably by binding to FKBP, but its molecular mechanism of action is different from that of FK–506 (98). The third component to which the RAP–FKBP complex binds has been recently identified, and is not Cn. RAP suppresses the phosphorylation and activity of the 70–kD S6 kinase (p 70^{s6k}) stimulated by IL–2 in T cells. This enzyme phosphorylates the ribosomal protein S6 on multiple serine residues, thereby increasing the efficiency of protein synthesis, which appears to be required in several of the steps of cell cycle progression. Its inhibition may therefore partly account for the antiproliferative effect of the drug. RAP also inhibits another cell–cycle controlling kinase, $p34^{cdc2}$, in IL–2–stimulated T cells. It may, therefore, alter multiple components of the signaling cascade that governs the response to cytokines in lymphoid and nonlymphoid cells.

Pharmacological concentrations of CsA (3 to 800 ng/ml) concentration-dependently inhibited histamine and LTC_4 release from basophils challenged with antigen or anti–IgE (30,95). The inhibitory effect was evident even when the drug was added during release. CsH, which has extremely low affinity for CyP, had no effect on activation of basophils by antigen and anti–IgE. CsA and its analogs inhibit the release of preformed and *de novo* synthesized mediators from basophils, presumably by interacting with CyP (95), because we found a significant correlation between the concentrations of CsA, CsG, CsC, and CsD that inhibited histamine release by 30% and their affinity for CyP. CsA, but not CsH, also inhibited the immunological release of histamine and LTC_4 and PGD_2 from human lung and skin mast cells, presumably by interacting with CyP (94,96,99).

In more recent experiments we demonstrated that a single oral dose of CsA (7 mg/kg) in normal volunteers caused rapid and significant inhibition of histamine release from basophils obtained *ex vivo* and challenged *in vitro* with anti–IgE, f–met–leu–phe (FMLP) and compound A23187 (33). The inhibitory effect of CsA was extremely rapid, peaking at 1–5 h then slowly declining up to 13 h by which time it was practically undetectable. This inhibition was associated with a sharp rise in the CsA blood level (max $\approx$ 500 ng/ml), that rapidly decreased within 5 h. Thus, the prompt inhibitory effect of CsA is correlated with rapid drug absorption and the effect persists up to 13 h, at which point CsA was undetectable (33).

In a second group of experiments, healthy volunteers were given oral CsA (5 mg/Kg) or placebo for five days. A constant plasma level of $\approx$ 250 ng/ml CsA was obtained. Basophil releasability in response to anti–IgE, FMLP and A23187 was diminished by 30 to 60 % throughout the course of CsA treatment. Placebo had no such effect. These experiments indicate that orally administered CsA rapidly inhibits histamine release from basophils obtained *ex vivo* and provide a rare example of how an *in vivo* administered drug can modulate basophil releasability *ex vivo*. These results and the additive inhibitory effect of combinations of CsA and corticosteroids on the release of mediators from basophils (66,100) might explain to some extent the efficacy of CsA in patients with severe corticosteroid-dependent asthma (85).

Compound FK-506 (1 to 300 n*M*) also inhibits histamine and LTC_4 release from basophils and mast cells challenged with antigen and anti-IgE, whereas RAP has little or no effect on IgE- and A23187-induced release. Interestingly, RAP does act as a competitive antagonist of FK-506, presumably at the level of FKBP (30,31,97). It thus appears that binding to FKBP is necessary, but not sufficient in itself to deliver the inhibitory signal for the release of proinflammatory mediators from basophils and mast cells.

CONCLUDING REMARKS

In this review we have touched on some recently elucidated aspects of mast cell and basophil immunopharmacology. It is now evident that in man the two cells types play different roles in inflammatory reactions, besides differing biochemically, immunologically and pharmacologically. Thanks to the advent of techniques for the isolation and purification of basophils from peripheral blood and mast cells from different anatomical sites (lung parenchyma, skin tissues, *etc.*) there is evidence that these cells differ in several aspects of the biochemical mechanisms underlying the signal transduction pathways. This heterogeneity of $Fc\varepsilon RI^+$ cells presumably reflects their different roles in various pathophysiological conditions.

Immunologically activated human mast cells and basophils can synthesize several cytokines (2-5) and proinflammatory mediators (1,25). Drugs modulating the secretion of proinflammatory mediators such as CsA and FK-506 might, at the same time, also affect the *in vitro* release of cytokines (101). It remains to be established whether this new class of drugs, with anti-inflammatory and immunosuppressive properties, has similar effects on human basophils and mast cells. If so, these compounds might play a dual role in the control of immune disorders by inhibiting the release of proinflammatory mediators and the *de novo* synthesis of immunoregulatory cytokines.

Efforts to gain further insight into the biochemical events occurring during the immunological activation of human mast cells and basophils should lead to the identification of new biochemical targets for pharmacological intervention. It is also becoming evident that the study of pharmacological control of basophils and possibly mast cells too should be extended to more complex *in vivo* systems. The results with CsA to date indicate that this model can also be applied to the *in vivo* modulation of human $Fc\varepsilon RI^+$ cells. These *in vitro* and *in vivo* results should eventually benefit individuals suffering not only from allergic diseases, but also from a variety of immunological disorders involving basophils and mast cells.

ACKNOWLEDGMENTS

The original work presented in this article was supported in part by grants from the C.N.R. (Project F.A.T.M.A.: Subproject Prevention and Control of Disease Factors; Project No. 94.00607.PF41) and the M.U.R.S.T. (Rome, Italy). The authors would like to express their special appreciation to Drs. R. Rickler and A. Renda for their kind help in providing the human lung and skin specimens that made possible some of the experiments described.

REFERENCES

1. Marone G, Casolaro V, Cirillo R, Stellato C, Genovese A. Pathophysiology of human basophils and mast cells in allergic disorders. *Clin Immunol Immunopathol* 1989;50:S24–S40.
2. Walsh LJ, Trinchieri G, Waldorf HA, Whitaker D. Human dermal mast cells contain and release tumor necrosis factor α, which induces endothelial leukocyte adhesion molecule 1. *Proc Natl Acad Sci USA* 1991;88:4220–4.
3. Bradding P, Feather IH, Howarth PH, Mueller R, Roberts JA, Britten K, Bews JPA, Hunt TC, Okayama Y, Heusser CH, Bullock GR, Church MK, Holgate ST. Interleukin 4 is localized to and released by human mast cells. *J. Exp. Med.* 1992;176:1381–6.
4. Möller A, Lippert U, Lessmann D, Kolde G, Hamann K, Welker P, Schadendorf D, Rosenbach, Luger T, Czarnetzki BM. Human mast cells produce IL-8. *J Immunol* 1993; 151:3261–6.
5. MacGlashan DW Jr, White JM, Huang S-K, Ono SJ, Schroeder JT, Lichtenstein LM. Secretion of IL-4 from human basophils. The relationship between IL-4 mRNA and protein in resting and stimulated basophils. *J Immunol* 1994; 152:3006–16.
6. Kimura I, Tanizaki Y, Saito K, Takahashi K, Ueda N, Sato S. Appearance of basophils in the sputum of patients with bronchial asthma. *Clin Allergy* 1975;5:95–8.
7. Pin I, Freitag AP, O'Byrne PM, Girgis-Gabardo A, Watson RM, Dolovich J, Denburg JA, Hargreave FE. Changes in the cellular profile of induced sputum after allergen-induced asthmatic responses. *Am Rev Respir Dis* 1992;145:1265–9.
8. Patterson R, McKenna JM, Suszko IM, Solliday NH, Pruzansky JJ, Roberts M, Kehoe TJ. Living histamine-containing cells from the bronchial lumens of humans: description and comparison of histamine content with cells of Rhesus monkeys. *J Clin Invest* 1977;59:217–25.
9. Tomioka M, Ida S, Shindoh Y, Ishihara T, Takishima T. Mast cells in bronchoalveolar lumen of patients with bronchial asthma. *Am Rev Respir Dis* 1984;129:1000–5.

10. Guo C-B, Liu MC, Galli SJ, Kagey-Sobotka A, Lichtenstein LM. The histamine containing cells in the late phase response in the lung are basophils. *J Allergy Clin Immunol* 1990;85:172 (abstract).
11. Liu MC, Hubbard WC, Proud D, Stealey BA, Galli SJ, Kagey-Sobotka A, Bleecker ER, Lichtenstein LM. Immediate and late inflammatory responses to ragweed antigen challenge of the peripheral airways in allergic asthmatics. *Am Rev Respir Dis* 1991;144:51-8.
12. Brinkman GL. The mast cell in normal human bronchus and lung. *J Ultrastructure Res* 1968;23:115-23.
13. Patterson R, Head LR, Suszko IM, Zeiss CRJr. Mast cells from human respiratory tissue and their *in vitro* reactivity. *Science* 1972;175:1012-4.
14. Crimi E, Chiaramondia M, Milanese M, Rossi GA, Brusasco V. Increase of mast cell numbers in bronchial mucosa after the late-phase asthmatic response to allergen. *Am Rev Respir Dis* 1991;144:1282-6.
15. Koshino T, Teshima S, Fukushima N, Takaishi T, Hirai K, Miyamoto Y, Arai Y, Sano Y, Ito K, Morita Y. Identification of basophils by immunohistochemistry in the airways of post-mortem cases of fatal asthma. *Clin Exp Allergy* 1993; 23:919-25.
16. Sparrow D, O'Connor T, Rosner B, Weiss TS. Predictors of longitudinal change in methacholine airways responsiveness among middle-aged and older men: the normative aging study. *Am J Respir Crit Care Med* 1994; 149:376-81.
17. Bédard PM, Brunet C, Pelletier G, Hébert J. Increased compound 48/80 induced local histamine release from nonlesional skin of patients with chronic urticaria. *J Allergy Clin Immunol* 1986;78:1121-5.
18. Casolaro V, Cirillo R, Genovese A, Formisano S, Ayala F, Marone G. Human basophil releasability. IV. Changes in basophil releasability in patients with chronic urticaria. *J Immunol Res* 1989;1:67-73.
19. Marone G, Giugliano R, Lembo G, Ayala F. Human basophil releasability.II. Changes in basophil releasability in patients with atopic dermatitis. *J Invest Dermatol* 1986;87:19-23.
20. Dvorak HF, Mihm MC. Basophilic leukocytes in allergic contact dermatitis. *J Exp Med* 1972;135:235-54.
21. Casolaro V, Galeone D, Giacummo A, Sanduzzi A, Melillo G, Marone G. Human basophil/mast cell releasability. V. Functional comparisons of cells obtained from peripheral blood, lung parenchyma and bronchoalveolar lavage in asthmatics. *Am Rev Respir Dis* 1989;139:1375-82.
22. Gauchat JF, Henchoz S, Mazzei G, Aubry JP, Brunner T, Blasei H, Life P, Talabot D, Flores-Romo L, Thompson J, Kishi K, Butterfield J, Dahinden C, Bonnefoy JY. Induction of human IgE synthesis in B cells by mast cells and basophils. *Nature* 1993;365:340-3.
23. Dvorak AM, Galli SJ, Schulman ES, Lichtenstein LM, Dvorak HF. Basophil and mast cell degranulation: ultrastructural analysis of

mechanisms of mediator release. *Fed Proc* 1983;42:2510-5.
24. Galli SJ: New concepts about the mast cell. *N Engl J Med* 1993;328:257-65.
25. Bascom R, Pipkorn U, Lichtenstein LM, Naclerio RM. The influx of inflammatory cells into nasal washings during the late response to antigen challenge. Effect of systemic steroid pretreatment. *Am Rev Respir Dis* 1988;138:406-12.
26. Marone G, Cirillo R, Genovese A, Marino O, Quattrin S. Human basophil/mast cell releasability. VII. Heterogeneity of the effect of adenosine on mediator secretion. *Life Sci* 1989;45:1745-54.
27. Naclerio RM, Proud D, Togias AG, Adkinson NFJr, Meyers DA, Kagey-Sobotka A, Plaut M, Norman PS, Lichtenstein LM. Inflammatory mediators in late antigen-induced rhinitis. *N Engl J Med* 1985;313:65-70.
28. Charlesworth EN, Hood AF, Soter NA, Kagey-Sobotka A, Norman PS, Lichtenstein LM. Cutaneous late-phase response to allergen: mediator release and inflammatory cell infiltration. *J Clin Invest* 1989;83:1519-26.
29. Iliopoulos O, Baroody FM, Naclerio RM, Bochner BS, Kagey-Sobotka A, Lichtenstein LM. Histamine-containing cells obtained from the nose hours after antigen challenge have functional and phenotypic characteristics of basophils. *J Immunol* 1992;148:2223-8.
30. de Paulis A, Cirillo R, Ciccarelli A, Condorelli M, Marone G. FK-506, a potent novel inhibitor of the release of proinflammatory mediators from human FcεRI$^+$ cells. *J Immunol* 1991;146;2374-81.
31. de Paulis A, Cirillo R, Ciccarelli A, de Crescenzo G, Oriente A, Marone G. Characterization of the anti-inflammatory effect of FK-506 on human mast cells. *J Immunol* 1991;147:4278-85.
32. Marone G. Immunosuppressive treatment of chronic asthma. In:Lichtenstein LM, Fauci AS. *Current Therapy in Allergy, Immunology, and Rheumatology*. Toronto: B.C. Decker; 1992:368-71.
33. Casolaro V, Spadaro G, Patella V, Marone G. *In vivo* characterization of the anti-inflammatory effect of cyclosporin A on human basophils. *J Immunol* 1993;151:5563-73.
34. Marone G, de Paulis A, Casolaro V, Ciccarelli A, Spadaro G, Patella V, Stellato C, Cirillo R, Genovese A. Are the anti-inflammatory properties of cyclosporin A useful in the treatment of chronic asthma? In:Melillo G, O'Byrne PM, Marone G. *Respiratory Allergy. Advances in Clinical Immunology and Pulmonary Medicine*. Amsterdam: Elsevier; 1993:251-60.
35. Lichtenstein LM, Margolis S. Histamine release *in vitro*. Inhibition by catecholamines and methylxanthines. *Science* 1968;161:902-3.
36. Chan SC, Li S-H, Hanifin JM. Increased interleukin-4 production by atopic mononuclear leukocytes correlates with increased cyclic adenosine

monophosphate-phosphodiesterase activity and is reversible by phosphodiesterase inhibition. *J Invest Dermatol* 1993;100:681-4.

37. Beavo, J.A., Reifsnyder, D.H. Primary sequence of cyclic nucleotide phosphodiesterase isozymes and the design of selective inhibitors. *Trends Pharmacol Sci* 1990;11:150-5.

38. Columbo M, Horowitz E, McKenzie-White J, Kagey-Sobotka A, Lichtenstein LM. Pharmacologic control of histamine release from human basophils induced by platelet activating factor. *Int Arch Allergy Immunol* 1993;102:383-90.

39. Peachell PT, Undem BJ, Schleimer RP, MacGlashan DWJr, Lichtenstein LM, Cieslinski LB, Torphy TJ. Preliminary identification and role of phosphodiesterase isozymes in human basophils. *J Immunol* 1992;148:2503-10.

40. Marone G, Columbo M, Triggiani M, Cirillo R, Genovese A, Formisano S. Inhibition of IgE-mediated release of histamine and peptide leukotriene from human basophils and mast cells by forskolin. *Biochem Pharmacol* 1987;36:13-20.

41. Torphy TJ, Livi GP, Balcarek JR, White FH, Chilton FH, Undem BJ. Therapeutic potential of isozyme-selective phosphodiesterase inhibitors in the treatment of asthma. *Adv Second Messenger Phosphoprotein Res* 1991:25;289-305.

42. Dent G, Giembycz MA, Rabe KF, Barnes PJ. Inhibition of eosinophil cyclic nucleotide PDE activity and opsonised zymosan-stimulated respiratory burst by type IV-selective PDE inhibitors. *Br J Pharmacol* 1991;103:1339-46.

43. Nielson CP, Vestal RE, Sturm RJ, Heaslip R. Effects of selective phosphodiesterase inhibitors on the polymorphonuclear leukocyte respiratory burst. *J Allergy Clin Immunol* 1990;86:801-8.

44. Balibrea-Cantero JL, Arias-Diaz J, Garcia C, Torres-Melero J, Simon C, Rodriguez JM, Vara E. Effect of pentoxifylline on the inhibition of surfactant synthesis induced by TNF-α in human type II pneumocytes. *Am J Respir Crit Care Med* 1994;149:699-706.

45. Yasui K, Komiyama A, Molski TFP, Sha'Afi RI. Pentoxyfilline and CD14 antibody additively inhibit priming of polymorphonuclear leukocytes for enhanced release of superoxide by lipopolysaccharide: possible mechanism of these actions. *Infection Immun* 1994;62:922-7.

46. Marone G, Ambrosio G, Bonaduce D, Genovese A, Triggiani M, Condorelli M. Inhibition of IgE-mediated histamine release from human basophils and mast cells by fenoterol. *Int Arch Allergy Appl Immunol* 1984;74:356-61.

47. Crane J, Pearce N, Flatt A, Burgess C, Jackson R, Kwong T, Ball M, Beasley R. Prescribed fenoterol and death from asthma in New Zealand, 1981-83: case-control study. *Lancet* 1989;1:917-22.

48. Spitzer WO, Suissa S, Ernst P, Horwitz RI, Habbick B, Cockroft D,

Boivin J-F, McNutt M, Buist AS, Rebuck AS. The use of β-agonists and the risk of death and near death from asthma. *N Engl J Med* 1992;326:501-6.
49. Nials AT, Ball DI, Butchers PR, Coleman RA, Humbles AA, Johonson M, Vardey CJ. Formoterol on airway smooth muscle and human lung mast cells: a comparison with salbutamol and salmeterol. *Eur J Pharmacol* 1994;251:127-35.
50. Stevenson DD. Cross-reactivity between aspirin and other drugs/dietary chemicals. A critical review. In:Pichler WJ, Stadler BM, Dahinden C, Pécoud AR, Frei PC, Schneider C, de Weck AL. *Progress in Allergy and Clinical Immunology.* Toronto: Hogrefe & Huber Publishers;1989:462-6.
51. Marone G, Kagey-Sobotka A, Lichtenstein LM. Effects of arachidonic acid and its metabolites on antigen-induced histamine release from human basophils *in vitro. J Immunol* 1979;123:1669-77.
52. Marone G, Kagey-Sobotka A, Lichtenstein LM. IgE-mediated histamine release from human basophils: differences between antigen E- and anti-IgE-induced secretion. *Int Arch Allergy Appl Immunol* 1981;65:339-48.
53. Marone G, Columbo M, Cirillo A, Condorelli M. Studies on the pathophysiology of aspirin idiosyncrasy. In:Bonomo L, Tursi A. *Recent Advances in Allergology*. Florence: O.I.C. Medical Press;1986:77-97.
54. Wojnar RJ, Hearn T, Starkweather S. Augmentation of allergic histamine release from human leukocytes by nonsteroidal anti-inflammatory analgesic agents. *J Allergy Clin Immunol* 1980;66:37-45.
55. Swingle KF, Moore GGI, Grant TJ. 4-nitro-2-phenoxymethanesulfonanilide (R-805): a chemically novel anti-inflammatory agent. *Arch Int Pharmacodyn* 1976;221:132-9.
56. Marino O, Casolaro V, Meliota S, Stellato C, Guidi G, Marone G. Inhibition of histamine release from human FcεRI⁺ cells by nimesulide. *Agents and Actions* 1992;32:C311-4.
57. Casolaro V, Meliota S, Marino O, Patella V, de Paulis A, Guidi G, Marone G. Nimesulide, a sulfonanilide non-steroidal antiinflammatory drug, inhibits mediator release from human basophils and mast cells. *J Pharmacol Exp Ther* 1993; 267:1375-85.
58. Andri L, Senna G, Betteli C, Givanni S, Scaricabarozzi I, Mezzelani P, Andri G: Tolerability of nimesulide in aspirin-sensitive patients. *Ann Allergy* 1994;72:29-32.
59. Meade EA, Smith WL, DeWitt DL. Differential inhibition of prostaglandin endoperoxide synthase (cyclooxygenase) isozymes by aspirin and other non-steroidal anti-inflammatory drugs. *J Biol Chem* 1993;268:6610-4.
60. Marone G, Casolaro V, Spadaro G, Genovese A. Bronchoalveolar lavage. In:Weiss EB, Stein M. *Bronchial Asthma. Mechanisms and Therapeutics.* Boston: Little Brown and Company Publishers;1993:309-

13.
61. Beasley R, Roche WR, Roberts JA, Holgate ST. Cellular events in the bronchi in mild asthma and after bronchial provocation. *Am Rev Respir Dis* 1989;139.806-17.
62. Azzawi M, Bradley B, Jeffery PK, Frew AJ, Wardlaw AJ, Knowles G, Assoufi B, Collins JV, Durham S, Kay AB. Identification of activated T lymphocytes and eosinophils in bronchial biopsies in stable atopic asthma. *Am Rev Respir Dis* 1990;142:1407-13.
63. Dunsky EH, Zweiman B, Fischler E, Levy DA. Early effects of corticosteroids on basophils, leukocyte histamine, and tissue histamine. *J Allergy Clin Immunol* 1979;63:426-32.
64. Lavker RM, Schechter NM. Cutaneous mast cell depletion results from topical corticosteroid usage. *J Immunol* 1985;135:2368-73.
65. Schleimer RP, Lichtenstein LM, Gillespie E. Inhibition of basophil histamine release by anti-inflammatory steroids. *Nature* 1981;292:454-5.
66. Stellato C, Casolaro V, Renda A, Genovese A, Marone G. Anti-inflammatory effect of deflazacort. *Int Arch Allergy Immunol* 1992;99:340-2.
67. Marone G, Stellato C, Renda A, Genovese A. Anti-inflammatory effects of glucocorticoids and cyclosporin A on human basophils. *Eur J Clin Pharmacol* 1993;45 (Suppl.1):S17-S20.
68. Schleimer RP, Schulman ES, MacGlashan DWJr, Peters SP, Hayes EC, Adams GK III, Lichtenstein LM, Adkinson NFJr. Effects of dexamethasone on mediator release from human lung fragments and purified human lung mast cells. *J Clin Invest* 1983;71:1830-5.
69. Pipkorn U, Proud D, Lichtenstein LM, Schleimer RP, Peters SP, Adkinson NFJr, Kagey-Sobotka A, Norman PS, Naclerio RM. Effect of short-term systemic glucocorticoid treatment on human nasal mediator release after antigen challenge. *J Clin Invest* 1987;80:957-61.
70. Laursen LC, Taudorf E, Weeke B: High-dose inhaled budesonide in treatment of severe steroid-dependent asthma. *Eur J Respir Dis* 1986;68:19-28.
71. Beaven MA, Rogers J, Moore JP, Hesketh TR, Smith GA, Metcalfe JC. The calcium signal and correlation with histamine release in 2H3 cells. *J Biol Chem* 1984;259.7129-36.
72. Warner JA, MacGlashan DWJr. Protein kinase C (PKC) changes in human basophils: IgE-mediated activation is accompanied by an increase in total PKC activity. *J Immunol* 1989;142:1669-77.
73. Nishizuka Y. The role of protein kinase C in cell surface signal transduction and tumor promotion. *Nature* 1984;308: 693-8.
74. White, KN, Metzger, H. Translocation of protein kinase C in rat basophilic leukemic cells induced by phorbol ester or by aggregation of IgE receptors. *J Immunol* 1988;141:942-7.
75. Schleimer RP, Gillespie E, Lichtenstein LM. Release of histamine from

human leukocytes stimulated with the tumor–promoting phorbol diesters. I. Characterization of the response. *J Immunol* 1981;126:570–4.

76. Patella V, Casolaro V, Ciccarelli A, Pettit GR, Columbo M, Marone G. The antineoplastic bryostatins affect differently human basophils and mast cells. *Blood*, 1994, *in press*.

77. Warner JA, MacGlashan DWJr. Signal transduction events in human basophils: a comparative study of the role of protein kinase C in basophils activated by anti–IgE antibody and formyl–methionyl–leucyl–phenylalanine. *J Immunol* 1990;145:1897–905.

78. Massey WA, Cohan VL, MacGlashan DWJr, Gittlen SW, Kagey-Sobotka A, Lichtenstein LM, Warner JA. Protein kinase C modulates immunoglobulin E–mediated activation of human mast cells from lung and skin. I. Pharmacologic inhibition. *J Pharmacol Exp Ther* 1991;258:824–9.

79. Coussens L, Parker PJ, Rhee L, Yang-Feng TL, Chen E, Waterfield MD, Francke U, Ullrich A. Multiple, distinct forms of bovine and human protein kinase C suggest diversity in cellular signaling pathways. *Science* 1986;233:859–66.

80. Bergstrand H, Lundquist B, Karabelas K, Michelsen P. Modulation of human basophil histamine release by protein kinase C inhibitors differs with secretagogue and with inhibitor. *J Pharmacol Exp Ther* 1992;260:1028–37.

81. Benhamou M, Gutkind JS, Robbins KC, Siraganian RP. Tyrosine phoshorylation coupled to IgE receptor mediated signal transduction and histamine release. *Proc Natl Acad Sci USA* 1990;87:5327–30.

82. Lavens SE, Peachell PT, Warner JA. Role of tyrosine kinases in IgE-mediated signal transduction in human lung mast cells and basophils. *Am J Respir Cell Mol Biol* 1992;7:637–44.

83. Borel JF. Comparative study of *in vitro* and *in vivo* drug effects on cell-mediated cytotoxicity. *Immunology* 1976;31:631–41.

84. Marone G, de Paulis A, Ciccarelli A, Casolaro V, Cirillo R. Mechanism(s) of action of cyclosporin A. *Seminars Clin Immunol* 1991;2:11–6.

85. Alexander AG, Barnes NC, Kay AB. Trial of cyclosporin in corticosteroid–dependent chronic severe asthma. *Lancet* 1992;339:324–8.

86. Hait WN, Harding MW, Handschumacher RE. Calmodulin, cyclophilin, and cyclosporin A. *Science* 1986;233:987–8.

87. Standaert RF, Galat A, Verdine GL, Schreiber SL. Molecular cloning and overexpression of the human FK506–binding protein FKBP. *Nature* 1990;346:671–4.

88. Bergsma DJ, Eder C, Gross M, Kersten H, Sylvester D, Appelbaum E, Cusimano D, Livi GP, McLaughlin MM, Kasyan K, Porter TG, Silverman C, Dunnington D, Hand A, Prichett WP, Bossard MJ, Brandt M, Levy MA. The cyclophilin multigene family of peptidyl–prolyl

isomerases. Characterization of three separate human isoforms. *J Biol Chem* 1991;266:23204-14.
89. Jin Y-J, Albers MW, Lane WS, Bierer BE, Schreiber SL, Burakoff SJ. Molecular cloning of a membrane-associated human FK506- and rapamycin-binding protein, FKBP-13. *Proc Natl Acad Sci USA* 1991;88:6677-81.
90. Takahashi N, Hayano T, Suzuki M. Peptidyl-prolyl cis-trans isomerase is the cyclosporin A-binding protein cyclophilin.*Nature* 1989;337:473-5.
91. Rosen MK, Standaert RF, Galat A, Nakatsuka M, Schreiber SL. Inhibition of FKBP rotamase activity by immunosuppressant FK506: twisted amide surrogate. *Science* 1990;248:863-6.
92. Liu J, Farmer JDJr, Lane WS, Friedman J, Weissman I, Schreiber SL. Calcineurin is a common target of cyclophilin-cyclosporin A and FKBP-FK506 complexes. *Cell* 1991;66:807-15.
93. O'Keefe SJ, Tamura J, Kincaid RL, Tocci MJ, O'Neil EA. FK-506- and CsA-sensitive activation of the interleukin-2 promoter by calcineurin. *Nature* 1992;357:692-4.
94. Marone G, Triggiani M, Cirillo R, Giacummo A, Siri L, Condorelli M. Cyclosporin A (CsA) inhibits the release of histamine and peptide leukotriene C_4 from human lung mast cells. *Ricerca Clin Lab* 1988;18:53-9.
95. Cirillo R, Triggiani M, Siri L, Ciccarelli A, Pettit GR, Condorelli M, Marone G. Cyclosporin A rapidly inhibits mediator release from human basophils presumably by interacting with cyclophilin. *J Immunol* 1990;144:3891-7.
96. Stellato C, de Paulis A, Ciccarelli A, Cirillo R, Patella V, Casolaro V, Marone G. Anti-inflammatory effect of cyclosporin A on human skin mast cells. *J Invest Dermatol* 1992;98:800-4.
97. de Paulis A, Stellato C, Cirillo R, Ciccarelli A, Oriente A, Marone G. Anti-inflammatory effect of FK-506 on human skin mast cells. *J Invest Dermatol* 1992;99:723-8.
98. Kuo CJ, Chung J, Fiorentino DF, Flanagan WM, Blenis J, Crabtree GR. Rapamycin selectively inhibits interleukin-2 activation of p70 S6 kinase. *Nature* 1992;358:70-3.
99. Triggiani M, Cirillo R, Lichtenstein LM, Marone G. Inhibition of histamine and prostaglandin D_2 release from human lung mast cells by cyclosporin A. *Int Arch Allergy Appl Immunol* 1989;88:253-5.
100. Marone G, de Paulis A, Ciccarelli A, Casolaro V, Spadaro G, Cirillo R. *In vitro* and *in vivo* characterization of the anti-inflammatory effects of cyclosporin A. *Int Archs Allergy Immun* 1992;99:279-83.
101. Hatfield SM, Mynderse JS, Roehm NW. Rapamycin and FK-506 differentially inhibit mast cell cytokine production and cytokine-induced proliferation and act as reciprocal antagonists. *J Pharmacol Exp Ther* 1992;261:970-6.

18

MAST CELL HYPERPLASIA AND ACTIVATION IN THE CONTEXT OF HELMINTH INFECTION, A ROLE FOR STEM CELL FACTOR?

Cheryl L. Scudamore*, George F. Newlands[#],
Stephen J. Galli[o] and Hugh R.P. Miller*

*Department of Veterinary Clinical Studies, University of Edinburgh,
[#]Moredun Research Institute, Edinburgh and
[o]Department of Pathology, Harvard Medical School, Boston

Mechanisms of immunity against helminth parasites in man are poorly understood despite the fact that infection is endemic in many parts of the world. Epidemiological studies indicate that age-dependent development of resistance to infection is due to acquired immunity. Where worm burden data are compared with immunological responses there is increasing evidence of a correlation between levels of parasite-specific IgE and development of resistance against a variety of helminth infections including, for example, *S. haematobium* in man (1). Similarly, recent investigations of peripheral blood T cell responses from patients infected with helminths suggest that a preponderance of cells secrete interleukins 4 and 5 following *in vitro* challenge with worm antigens (2), thus further implicating IgE, as well as eosinophils, in the response to helminths in man.

The correlations between IgE and development of resistance have been most intensively investigated in laboratory animals. Direct evidence implicating IgE in resistance in rats to the nematode *Trichinella spiralis* was provided by the passive transfer of parasite-specific IgE (3). However, this treatment was only effective when preceded by the adoptive transfer of immune lymphocytes. The combined effect of parasite-specific IgE and of immune cells was to confer significant resistance to challenge with infective larvae. About 50% of the latter were rapidly expelled within 24 hours (3). This phenomenon of rapid expulsion (RE) has been described in a number of species (4) but there is, as yet, no comparable data in man.

Three cell types are most closely associated with IgE-mediated responses against helminths; eosinophils, basophils and mast cells (4). The cell most likely to be involved in RE is the mast cell since, once developed, it remains resident in the tissues for several weeks whereas basophils and eosinophils, survive for relatively short periods (4). There is now increasing evidence that the mucosal mast cell (MMC) does play a role in the rejection of certain intestinal nematodes (4). The purpose of this paper is to review that evidence and to present new data on mast cell activation which may throw light on the functional role of MMC in helminthosis. Since massive hyperplasia of mast cells occurs during infection, the involvement in this process of stem cell factor (SCF), the ligand for the tyrosine kinase membrane receptor c-*kit*, abundant on mast cells (5), is also reviewed.

MAST CELL HYPERPLASIA AND MAST CELL GRANULE CHYMASES IN THE PARASITIZED HOST

In the field, exposure of the at-risk population to helminth infection can vary from a low intensity, sporadic challenge to an almost daily intake of large numbers of infective larvae. An example of intense exposure is provided by food animals grazing permanent pasture or housed under conditions of poor hygiene. It is not unreasonable, for example, for a sheep or cow to ingest thousands of nematode larvae/day and it is quite clear that the majority of adult animals, where nutrition is adequate and intercurrent disease is absent, are highly resistant to reinfection (4). As yet, the picture in man is less clear because there is insufficient epidemiological data comparing the level of challenge with the level of infection.

The first real evidence that allergic reactions were responsible for the manifestation of resistance to infection came from studies in sheep. It was noted that a sudden intake of larvae caused the expulsion of the existing adult worm population and this was associated with increased levels of histamine in the blood (6). Subsequent studies in rats suggested that intestinal anaphylaxis, in association with passively transferred antibodies, was sufficient to promote worm expulsion (7). These functional studies suggested, indirectly, that mast cells and basophils could be involved in mucosal allergic reactions against helminth infection.

Mast cell hyperplasia is a predominant feature of long term or repeated helminth infections (4, 6). Increased numbers of mast cells have been reported in nematode-infected human intestine but it is in the rat, mouse, guinea pig and sheep where helminth-associated mast cell hyperplasia has been most intensively investigated (4, 6). Early data relied on tissue counts to show the association between mast cells and helminthosis and, although much work was done to examine the roles of histamine and serotonin (6), it was always difficult to determine whether these 2 mediators were mast cell-derived or originated from basophils or enteroendocrine cells (4).

Woodbury *et al* (8) first reported that the soluble granule chymase, rat mast cell protease-II (RMCP-II) was present in intestinal mucosal mast cells (MMC) and it was subsequently noted that the highly insoluble granule chymase RMCP-I was absent from MMC but present in connective tissue mast cells (CTMC) (9). In fact, the distribution of the two mast cell granule chymases was virtually mutually exclusive (9). Importantly, Woodbury and colleagues had developed immunoassays to quantify RMCP-II and it was thus possible to show that mucosal mastocytosis was associated with a massive increase in the tissue content of RMCP-II (Table I) (4, 10).

A murine intestinal mast cell granule chymase, mouse mast cell protease-1 (MMCP-1), and a mast cell granule chymase from sheep (SMCP) have been purified and characterized biochemically (11, 12). Antibodies raised against these 2 proteases have confirmed that both are present in MMC and immunoassays have shown that concentrations of MMCP-1 and SMCP are substantially increased in helminth infection (4). As yet there have been no comparable studies in man.

Table I Mast cell recruitment and activation in parasitized rats

PARASITE	TISSUE	TISSUE RMCP-II (μg/gm)	SYSTEMIC RMCP-II (ng/ml)
N BRASILIENSIS	JEJUNUM	3,000	1000-2000
T SPIRALIS	JEJUNUM	3,000	>4,000
M CORTI	LIVER	400	<100
S MANSONI	LIVER	297	1000
P CROTALI	MESENTERY	N/A	100
NONE	JEJUNUM	300	250
	LIVER	0.9	
N.b Antigen	INTRADUODENAL	3,000	10^4
N.b Antigen	INTRAVENOUS	N/A	10^5-10^6

The range of parasites, their location in the tissues and the concentrations of RMCP-II in the tissues after infection are shown in this table (reviewed in 4, 13 and 14). Also included are the values for control, non-infected rats, and for rats primed by previous infection with N. brasiliensis and challenged with worm antigen (N.b Antigen). The maximal systemic release of RMCP-II in response to infection or challenge is also shown. It is particularly interesting to note that infection with P. crotali recruits MMC but suppresses the secretion of RMCP-II (15); the same is probably true for M. corti, but the ELISA used for this study was not sufficiently sensitive to detect systemic RMCP-II (16).

RMCP-II, MMCP-1 and SMCP predominate in MMC and it is estimated that 94% of total RMCP-II and >99% of total MMCP-1 are present in gastrointestinal MMC (4). Therefore, the observation that these proteases are secreted systemically during helminth infection and during intestinal anaphylaxis (Table I) provides a unique opportunity to monitor and quantify the activation of MMC *in vivo*.

Table II Baseline values for mouse mast cell protease-1 in different strains of mice

STRAIN	JEJUNUM ng/gm wet weight	SERUM ng/ml
NIH	2000	200
(CBA x BALB/C F_1	137	11
(WB x $C_{57}B_L/6$) F_1- +/+	90	0.2
(WB x $C_{57}B_L/6$) F_1- W/W^v	40	0

MUCOSAL MAST CELL ACTIVATION *IN VIVO*

a) Baseline Secretion Of Mast Cell Proteases.

The development of more sensitive enzyme-linked immunoassays has allowed a detailed analysis of baseline blood values for MMCP-1 and RMCP-II (Tables I and II) but not for SMCP. In the latter instance, unidentified plasma inhibitors block the binding of antibody to SMCP (14). By contrast, although both MMCP-1 and RMCP-II are inhibited by alpha-1-proteinase inhibitor (k_{ass}, 2 to 12 x 10^5 M^{-1} s^{-1}) (14) the proteinase-inhibitor complexes are readily quantified by ELISA (14). The baseline values for MMCP-1 in different mouse strains show considerable variation with the highest values for serum and for intestinal homogenates in NIH mice and the lowest in mast cell deficient W/W^v mice (14) (Table II). The baseline values in rats vary between 70 and 300 ng RMCP-II/ml of serum (14) but this variability has not been assessed in terms of rat strain. In the intestine, there usually are 200-400 µg RMCP-II/gm wet weight of jejunal homogenate (14) which is several orders of magnitude greater than the MMCP-1 content of mouse jejunum (14). Several lines of evidence indicate that the baseline levels of MMC proteases in the blood of rats and mice reflect secretion, rather than the appearance of proteases derived from dead or damaged MMC (4). For example, dexamethasone treatment markedly diminishes numbers of MMC, but does not result in a significant associated increase in levels of circulating proteases (4).

The stimulus for the continued secretion of MMC proteases into blood in apparently normal laboratory rats and mice has not been identified. Whilst the data in Table II suggest that the differences could be strain-related it is also possible that disease status or other physiological factors associated with different housing conditions affect the baseline secretion of RMCP-II and MMCP-1. It is highly likely that systemic secretion represents a continuous process since the proteinase-inhibitor complexes formed in the blood will be rapidly cleared by the liver and there must, therefore, be a steady-state where the rate of secretion is balanced by the rate of clearance.

One likely stimulus for secretion is through the local release of cytokines in the gut since it has been shown that daily treatment of C57BL/6 mice with recombinant IL-3 was associated with a 4-5 fold increase in the concentration of MMCP-1 in serum (17). There was also a substantially increased level in the gut lumen which was associated with recruitment of MMC in the mucosa

(17). However, it is important to note that the main effect of IL-3 was to increase the number of mast cells (17).

It has also been shown that the systemic secretion of RMCP-II can be depressed by irradiation and cyclosporin-A treatments which are both associated with depletion of MMC (4). Baseline RMCP-II values were also significantly depressed in rats harbouring the pentastomid parasite *Porocephalus crotali* (15). The rat is an intermediate host for this parasite whose definitive host is the rattlesnake and the nymph stages survive many months in cyst-like granulomas which are heavily populated with MMC and can contain up to 90 µg RMCP-II/cyst (15). It is possible, for example, that the parasite secretes serpin-like molecules, thereby reducing the levels of detectable RMCP-II. Alternatively it may down-regulate cytokine levels in the host.

b) Enhanced Secretion Of Mast Cell Proteases During Helminthosis

The systemic release of mast cell granule chymases was demonstrated for the first time in rats during the RE of *N. brasiliensis* from the jejunum (4) and was subsequently recorded in parasitized mice and sheep (14). In all three species, maximal secretion was coincident with worm expulsion from the gut. However, it was also evident in rats and mice that *T. spiralis* stimulated higher levels of systemic secretion than did *N. brasiliensis* or other helminth parasites (Table I). In both mice and sheep, systemic secretion during infection with GI nematodes was associated with local secretion of chymases into the gut lumen (14). Comparable studies have not been done in the rat.

Systemic secretion of chymase in rats infected with *N. brasiliensis* reaches a maximum at the time of worm expulsion but precedes, by several days, the maximum accumulation of mast cells and RMCP-II in the jejunum (14). In contrast, in both rats and mice, the immune expulsion of *T. spiralis* coincides with maximal chymase secretion and maximal accumulation of mast cells and chymases in the mucosa (14). In neither instance has the stimulus responsible for secretion been identified. It is possible that sufficient parasite-specific IgE is produced to sensitize MMC prior to worm expulsion even though none is detected in serum (7). Alternatively, the secretory response of MMC may be upregulated by cytokines (17) and this possibility is discussed below.

c) Analysis Of The Systemic Release Of RMCP-II And Mucosal Permeability By Ex-Vivo Perfusion.

It has long been hypothesised that mast cells are involved in alterations in mucosal permeability during helminth infection (4). This hypothesis is supported by ultrastructural analysis, by *in vivo* studies of permeability changes in relation to the systemic secretion of RMCP-II, and by measurement across the mucosa of short circuit current changes *in vitro* (reviewed in 4). However, it is difficult to quantify the relationship between mast cell activation and changes in permeability *in vivo* sufficiently precisely. Therefore the vasculature of the rat jejunum was perfused *ex vivo* via the cranial mesenteric artery with Krebs Ringer solution, gassed with 95% O_2 and 5% CO_2, and containing 10% Ficoll 70 and 10 mg/ml Human serum albumin/Evan's blue (HSA/EB). The lumen of a 25-30 cm segment of jejunum was also perfused with Krebs Ringer. The temperature of the perfused intestine was maintained at 37° C throughout. Vascular perfusate outflow was collected via the portal vein having first occluded the splenic, gastric, and distal intestinal blood vessels so that the perfusate was contained within the jejunal vasculature. Samples were collected at 2 minute intervals prior to challenge and at 1 minute intervals thereafter.

Three groups of rats (Male Wistars, 230-300 gm) were used. Rats in Group I (n=5) were normal uninfected controls. Rats in Groups II (n=3) and III (n=5) had experienced infection with *N. brasiliensis* (3,000 L_3) 4 and 1 week previously. Challenge was with BSA (Group II) or with soluble adult *N. brasiliensis* worm homogenate (Groups I and III). Preliminary studies had shown that 100 worm equivalents was an optimum challenge dose.

The baseline output of RMCP-II into the gut lumen or into the vascular perfusate was significantly greater in the immune controls (Group II) than in the uninfected control group, but there was no significant change in the amount of RMCP-II or HSA/EB reaching the lumenal perfusate after challenge in either group. However, in the primed rats (Group III) there was a significant increase in the concentration of RMCP-II in the vascular perfusate within 70 seconds of challenge and maximal levels (~10 µg/ml) were achieved 1 minute later. The leak of RMCP-II into the gut lumen followed very similar kinetics except that it was delayed by a further 2 minutes. Similarly, the translocation of HSA/EB to the gut lumen was apparently delayed with maximal levels occurring at least 2 minutes after the first significant vascular leak of RMCP-II was detected.

The overall results are summarized in Fig 1 which, while it does not detail the kinetics of the changes, shows that very substantial quantities of RMCP-II are released into both the vascular bed and the gut lumen of primed rats and that this is associated with increased mucosal permeability to macromolecules. This mechanism is likely to be mast cell-mediated because:- a) the response is so rapid b) RMCP-II is released in substantial amounts c) circulating factors such as complement and immunoglobulins have been removed by perfusion and d) there is a body of data showing that activation of MMC is a feature of systemic anaphylaxis (4, 18, 19).

Recent experiments in which 1.5 mg of RMCP-II (100% active) has been infused into the vascular perfusate (100 µg RMCP-II/ml perfusate) resulted in the almost immediate translocation of RMCP-II and HSA/EB into the gut lumen. However, despite the translocation of microgramme quantities of RMCP-II into the gut lumen, there was no evidence of severe damage to the mucosa regardless of whether the chymase was released from the mast cell or was perfused via the vasculature. Histological examination has so far revealed minimal pathology and only in one or two instances was it possible to detect any loss of epithelial cells.

These results suggest that RMCP-II, despite its abundance in parasitized jejunal mucosa (Table I) may have a more subtle role than hitherto suspected. Previous *in vivo* experiments established that up to 1 mg of RMCP-II was released into the gut lumen within 5 minutes of systemic challenge and this was associated with massive shedding of the epithelium (18). Furthermore, *in vivo*, the systemic accumulation of RMCP-II was a much slower process, reaching a maximum of 1-2 mg/ml 30-60 minutes after challenge (18). However, recent studies have shown during systemic anaphylaxis that there is a substantial drop in blood flow to the mesentery with severe vascular congestion and probable mucosal anoxia (19, 20). If it was assumed that only

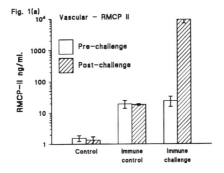
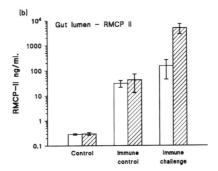
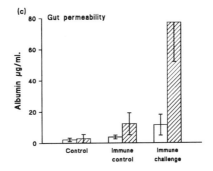

Figure 1:- Overall mean ± SEM concentrations of RMCP-II released into vascular perfusate (a) and gut lumen (b) from mucosal mast cells and Evans blue labelled albumin translocated from the vascular perfusate into the gut lumen (c). The pre-challenge period (open bars) represents the first 20 minutes of perfusion used to establish basal levels of RMCP-II release. The post-challenge period (shaded bars) represents the 10 minutes following antigen infusion.

10% of the recruited MMC were activated during systemic anaphylaxis then 300 µg of RMCP-II would be released into each gramme wet weight of jejunum and concentrations of RMCP-II within the interstitial tissues might be in the order of several milligrammes/ml. This is not the case in the perfusion system described here because, although flow rate decreases by some 15% within 2 minutes of antigen challenge in primed animals, it is still probably sufficient to rapidly clear the interstitial pool of RMCP-II.

The data obtained by perfusion are in good agreement with studies by Woodbury *et al* (21) which suggested that RMCP-II and human cathepsin G both caused increased permeability in cultured pulmonary epithelium without causing cell loss or cell death. As little as 100 ng RMCP-II/ml was sufficient to cause a 6-fold increase in epithelial permeability.

Although systemic anaphylaxis is a rather severe example of mucosal damage, the mechanisms involved may be representative of the type of lesion occurring during the recruitment of MMC response where, for example, in *T. spiralis* infection, 4-5 µg RMCP-II/ml blood are detected at the time of worm expulsion (Table I). Similarly, the work of Castro and colleagues (22) provided evidence that mucosal mastocytosis is an essential component of the early phase of a biphasic chloride secretory response *in vitro* when primed rat jejunum was challenged with worm antigen. Perdue and colleagues (23) showed that the intestinal tissues of genetically mast cell-deficient mice exhibited a markedly diminished early phase of the biphasic chloride secretory response to antigen challenge *in vitro* and that this defect was repaired in W/W^v mice which had undergone reconstitution of intestinal mast cell populations by transplantation of bone marrow cells from the congenic normal (+/+) mice.

Overall, the *ex vivo* perfusion has clearly established that RMCP-II and albumin are translocated across the epithelium within one or two minutes of anaphylactic activation of MMC. Furthermore, the studies where RMCP-II was perfused into the vasculature indicate either that the protease is directly responsible for altering mucosal permeability or that it triggers the release of permeability-inducing mediators from cells resident in the mucosa.

STEM CELL FACTOR AND C-*KIT* IN HELMINTHOSIS

a) c-*kit* And Stem Cell Factor Mutants

The importance of c-*kit* and of its ligand, stem cell factor, in the recruitment, proliferation and activation of mast cells is discussed by Drs Galli, Kitamura and Nakahata in this volume. In retrospect, much of our current understanding of the role of these two molecules in mast cell development was suggested by the original reports by Kitamura and colleagues that W/W^v and Sl/Sl^d mutant mice were mast cell-deficient, and by the subsequent work with these mutant mice (reviewed in 24, 25).

The use of these mutant mice to study helminth-induced mast cell hyperplasia strongly suggests that, for certain intestinal nematodes, mast cells do play a significant role in the rejection process. The results of these experiments, which have been summarized in several recent reviews (4, 24, 26), suggest a more important role for mast cells in the immune rejection of *T. spiralis* and *S. ratti* than in the response to *N. brasiliensis*. Importantly, repopulation of W/W^v mice with bone marrow cells from congenic +/+ mice 6 weeks before challenge was associated with mucosal mast cell hyperplasia and the normalization rate of expulsion of *T. spiralis* and *S. ratti* by these mutant mice (24, 26).

b) c-*kit* And Stem Cell Factor *In Vivo*

The phenotypic abnormalities expressed by mutant W/W^v and Sl/Sl^d mice were suggestive of several roles for SCF and c-*kit* involving regulation of the migration, maturation and proliferation of mast cells (24, 25, 27). There is also evidence that signalling through the c-*kit* receptor activates mast cells to secrete mediators (25, 27, 28). This was emphasized in more recent studies where rat bone marrow-derived mast cells (BMMC), which, in many characteristics, resemble rat MMC, secreted large amounts of RMCP-II into the supernatant when cultured in the presence of rrSCF (29). One of the questions raised by this and other work on the secretory response induced by SCF is whether this cytokine can play a similar role *in vivo*.

There are now several lines of evidence to suggest that c-*kit* is involved in the regulation of mast cell hyperplasia during intestinal nematodosis. The most recent evidence supporting a role for c-*kit* and, by implication, its ligand came from a study by Grencis *et al* (30) where treatment of mice infected with *T. spiralis* with anti c-*kit* antibody abrogated both mucosal mast cell hyperplasia and systemic secretion of MMCP-1. This antibody also suppressed worm expulsion, again suggesting that mucosal mast cell hyperplasia is a significant component of the rejection process. Huff and colleagues (31) have shown that mast cell progenitors are recruited to mesenteric lymph nodes in parasitized mice and express c-*kit* and that antisense oligonucleotides to c-*kit* block mast cell colony formation in the presence of fibroblast-conditioned medium (30). Since the fibroblast-conditioned medium probably contained SCF it was concluded that progenitor differentiation was SCF-dependent. It was, however, established that IL-3 could, similarly, support the growth of mast cell colonies and a dual cytokine signalling pathway for mast cell development was proposed (31).

The question raised by these data is, what is the relative importance in mast cell recruitment of SCF, a fibroblast-derived cytokine, versus IL-3 which is produced by T cells? Since there is ample evidence to show that mucosal mast cell hyperplasia is T-cell dependent and that IL-3 and IL-4 are involved in the *in vivo* recruitment of MMC in parasitized mice (4), the role of SCF remains open to question. The second point is whether SCF plays a significant role in upregulating the secretion of chymases during the development of mucosal mast cell hyperplasia.

c) The Role of Rat Stem Cell Factor In Nippostrongylosis

Two approaches were used to explore the functional role of SCF during infection, the first was to inject $rrSCF^{164}$ into rats infected with *N. brasiliensis* and to monitor the development of mast cells in different tissues and to quantify the systemic secretion of RMCP-II. The second was to raise polyclonal sheep antibodies against $rrSCF^{164}$ and to inject the affinity purified antibodies into *Nippostrongylus*-infected rats.

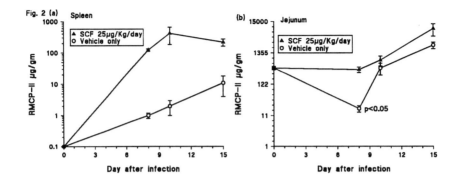

Figure 2:- The RMCP-II content of (a) spleen and (b) jejunum in Nippostrongylus infected rats is altered by administration of rr SCF^{164} at 25μg/kg/day for 14 days. Levels are significantly increased (P<0.001) in spleen at all time points during infection. In contrast, rr SCF^{164} has had a significant effect only on day 8 of infection in the jejunum.

Table III

**Administration of rr SCF^{164} during Nippostrongylosis.
Mast cell responses in jejunum compared with responses in spleen and liver.**

MAST CELLS / 0.2 mm^2 on day 15

INFECTION	SCF	JEJUNUM	SPLEEN	LIVER
-	-	26 ± 1	0 ± 0	1 ± 0.2
-	+	49 ± 4	70 ± 7	19 ± 3
+	-	238 ± 27	5 ± 2	4 ± 0
+	+	221 ± 40	179 ± 16	38 ± 7

The administration of rrSCF^{164} (25μg/Kg/day for 14 days) (32) into rats infected with 3000 L$_3$ on day 0 had, at best, a marginal effect on the development of mucosal mast cells and this was evident on day 8 when treated infected rats had a significant (P<0.05) increase in the jejunal content of RMCP-II when compared with saline-treated infected controls (Fig 2). There was, however, no other effect on intestinal mucosal mast cell development in the infected rats as judged by mast cell counts (Table III) nor by measuring the RMCP-II content of the jejunum (Fig 2). Similarly, the only difference in the level of systemic RMCP-II between saline and rrSCF-treated infected rats was on day 8 where a significant (P<0.05) 3-fold increase was found in treated rats. Continuous treatment with rrSCF^{164} for 14 days had no effect on the basal secretion of RMCP-II in uninfected control rats (data not shown) despite the fact that mast cell numbers in the jejunal mucosa increased two-fold (Table III).

The major effect of rrSCF^{164} was, as we have reported previously (32), to substantially increase mast cell development in other organs including spleen (Fig 2, Table III) and liver (Table III). The combination of infection with *N. brasiliensis* and treatment with rrSCF^{164} promoted a particularly intense mast cell response in the spleen (Table III) where concentrations of the connective tissue mast cell chymase, RMCP-I, increased from 3 ± 0.4 μg/gm in infected controls to 2250 ± 211 μg/gm in SCF^{164} treated rats on day 15 of infection. This compares with the much lower RMCP-II values in spleen (Fig 2), which suggests that a preponderance of CTMC-like cells developed in the spleen.

Treatment with rrSCF^{164} thus has a profound effect on mast cell development in some sites during infection but not in the jejunum where the worms reside. One possible explanation is that SCF levels in the mucosa are already sufficient to permit mast cell hyperplasia in this setting, which may, importantly, reflect the actions of additional cytokines such as IL-3 and/or IL/4 (4). The likely explanation for the widespread development of mast cells in non mucosal tissues is that progenitor cells circulate to many tissues during infection but normally are not retained in and/or differentiate at these sites. The administration of exogenous SCF^{164} may permit the retention and/or differentiation of these progenitors. Importantly, there was only a marginal effect of exogenous SCF^{164} on the systemic secretion of RMCP-II and that was associated with increased levels of RMCP-II in the jejunal mucosa on day 8 of infection. This suggests that secretion of mediators by MMC may be stimulated by other factors.

Experiments have also been completed using polyclonal sheep anti-rrSCF^{164}. Affinity purified antibody (1 mg/day) administered on days 3, 5, 7, 10 and 12 of infection with 3,000 *N. brasiliensis* profoundly suppressed (P<0.001) mucosal mast cell hyperplasia on day 10. The levels of jejunal RMCP-II and the systemic secretion of RMCP-II were, similarly, suppressed on days 6 and 10 of infection. These results support a role for SCF in the mucosal mast cell hyperplasia which occurs during *N. brasiliensis* infection.

CONCLUSIONS

Mast cell hyperplasia in association with helminth infection has been reported in a number of species including man (4, 33). When it has been possible to compare worm burden kinetics with mast cell development, numerous reports indicate either that worm expulsion precedes the mast cell hyperplasia or that mast cells are present while worm burdens remain stable (4). However, it is now clear that the simple enumeration of mast cells may bear little or no relationship to the extent to which the mast cell population as a whole is either activated or quiescent. The quantification of granule chymases has clearly established that, in normal uninfected rodents, there are significant levels of proteases derived from enteric mast cells in the peripheral blood, and several lines of evidence indicate that this reflects the secretion of these proteases by enteric mast cells (4). The *ex vivo* perfusion system described here reveals, for the first time, that RMCP-II can be secreted both into the vasculature and into the gut lumen in normal rats and that this basal secretion is substantially increased in rats primed by previous infection. The anaphylactic release of RMCP-II in the perfused gut is closely associated with increased mucosal permeability and our results, therefore, support the hypothesis that RMCP-II and, perhaps, other mast cell chymases, are responsible for modulating enteric epithelial permeability. Importantly, increased permeability can occur with minimal morphological changes in the mucosa. This suggests that the native substrate targets for RMCP-II may be limited, and that the mucosal damage which is observed in association with enteric mast cell activation *in vivo* may reflect, at least in part, the recruitment of additional effector mechanisms.

Whilst it is probable that the anaphylactic release of RMCP-II, is mediated by mast cell-bound IgE (4), the stimuli promoting basal secretion, or the early secretion during primary infection are not known. SCF promotes the secretion of RMCP-II by cultured BMMC (29) but there was little evidence to suggest that it has the same effect *in vivo*. However, SCF appears to be necessary for survival of MMC in both normal and parasitized rats. Since SCF can induce mast cell hyperplasia in experimental primates and humans (25), there is a strong possibility that, in man, SCF contributes to the development of mast cells during parasitic infection. One of the more intriguing questions now to be answered, given the T cell-dependency of mucosal mast cell hyperplasia, is whether the expression of SCF in parasitized tissues might also be influenced by T cell-dependent mechanisams.

ACKNOWLEDGMENTS

This work was supported by grants from the Wellcome Trust and the Scottish Office Agriculture and Fisheries Department, the United States Public Health Service (AI 22674 and AI 23990) and AMGEN, Inc.

REFERENCES

1. Hagan P, Blumental UJ, Dunn D, Simpson AJG, Wilkins HA. Human IgE, IgG$_4$ and resistance to reinfection with *Schistosoma haematobium*. *Nature* 1991;349:243-5
2. Mahanty S, Abrams JS, King CL, Limaye AP, Nutman TB. Parallel regulation of IL-4 and IL-5 in human helminth infections. *J. Immunol* 1992;148:3567-71.

3. Ahmad A, Wang CH, Bell RG. A role for IgE in intestinal immunity. Expression of rapid expulsion of *Trichinella spiralis* in rats transfused with IgE and thoracic duct lymphocytes. *J Immunol* 1991;146:3563-70.
4. Miller HRP. Mast cells: their function and heterogeneity. In: Moqbel R. *Allergy and Immunity to helminths: common mechanisms or divergent pathways?* London: Taylor and Francis;1992:228-48.
5. Galli SJ, Geissler EN, Wershil BK, Gordon JR, Tsai M, Hammel I. Insights into mast cell development and function derived from analyses of mice carrying mutations at *beige*, *W/c-kit* or *Sl/SCF* (*c-kit* ligand) loci. In: Kaliner MA, Metcalfe DD. *The Role of the Mast Cell in Health and Disease.* New York: Marcel Dekker;1992:129-202.
6. Rothwell TLW. Immune expulsion of parasitic nematodes from the alimentary tract. *Int J Parasit* 1989;19:139-68.
7. Jarrett EEE, Miller HRP. Production and activities of IgE in helminth infection. *Prog Allergy* 1982;31:178-233.
8. Woodbury RG, Gruzenski GM, Lagunoff D. Immunofluorescent localization of a serine protease in rat small intestine. *Proc Natl Acad Sci USA* 1978;75:2785-89.
9. Gibson S, Miller HRP. Mast cell subsets in the rat distinguished immunohistochemically by their content of serine proteinases. *Immunology* 1986;58:101-4.
10. Woodbury RG, Miller HRP. Quantitative analysis of mucosal mast cell protease in the intestines of *Nippostrongylus*-infected rats. *Immunology* 1982; 46:487-91.
11. Newlands GFJ, Gibson S, Knox DP, Grencis R, Wakelin D, Miller HRP. Characterization and mast cell origin of a chymotrypsin-like proteinase isolated from intestines of mice infected with *Trichinella spiralis*. *Immunology* 1987;62:629-34.
12. Huntley JF, Gibson S, Knox D, Miller HRP. The isolation and purification of a proteinase with chyymotrypsin-like properties from ovine mucosal mast cells. *Int J Biochem* 1986;18:673-82.
13. Miller HRP, Huntley JF, Newlands GFJ, Irvine J. Granule chymases and the characterization of mast cell phenotype and function in rat and mouse. *Monogr Allergy* 1990;27:1-30.
14. Miller HRP, Huntley JF, Newlands GFJ. Mast cell chymases in helminthosis and hypersensitivity. In: Caughey GH. *The Biology of Mast Cell Proteases.* New York: Marcel Dekker; 1994 (in press).
15. McHardy P, Riley J, Huntley JF. The recruitment of mast cells exclusively of the mucosal phenotype into granulomatous lesions caused by the pentastomid parasite *Porocephalus crotali*: Recruitment is irrespective of site. *Parasitology* 1993;106:47-54.
16. Chernin J, Miller HRP, Newlands GFJ, McLaren DJ. Proteinase phenotypes and fixation properties of rat mast cells in parasitic lesions caused by *Mesocestoides corti*: selective and site-specific recruitment of mast cell subsets. *Parasite Immunol* 1988;10:433-42.
17. Abe T, Sugaya H, Ishida K, Khan WI, Tasdemir I, Yoshimura K. Intestinal protection against *Strongyloides ratti* and mastocytosis induced by administration of interleukin-3 in mice. *Immunology* 1993;80:116-21.
18. King SJ, Miller HRP. Anaphylactic release of mucosal mast cell protease and its relationship to gut permeability in *Nippostrongylus*-primed rats. *Immunology* 1984;51:653-60.

19. Ramaswamy K, Mathison R, Carter L, Kirk D, Green F, Davison JS, Befus D. Marked antiinflammatory effects of decentralization of the superior cervical ganglia. *J Exp Med* 1990;172:1819-30.
20. Mion F, Cuber J-C, Minaire Y, Chayvialle J-A. Short term effects of indomethacin on rat small intestinal permeability. Role of eicosanoids and platelet activating factor. *Gut* 1994;34:390-5.
21. Woodbury RG, Le Trong H, Cole K, Neurath H, Miller HRP. Rat mast cell proteases. In: Galli SJ, Austen KF. *Mast cell and basophil differentiation and function in health and disease.* New York: Raven Press;1989:71-9.
22. Harari Y, Castro GA. Stimulation of parasite-induced gut hypersensitivity: implications for vaccination. *Immunology* 1989;66:302-7.
23. Perdue MH, Masson S, Wershil BK, Galli SJ. Role of mast cells in ion transport abnormalities associated with intestinal anaphylaxis. Correction of the diminished secretory response in genetically mast cell-deficient W/W^v mice by bone marrow transplantation. *J Clin Invest* 1991;87:687-93.
24. Kitamura Y, Kasugai T, Nomura S, Matsuda H. Development of mast cells and basophils. In: Foreman J. *Immunopharmacology of mast cells and basophils.* London: Academic Press 1993:5-27.
25. Galli SJ, Zsebo KM, Geissler EN. The kit ligand, stem cell factor. *Adv Immunol* 1994;55:1-96.
26. Reed ND. Function and regulation of mast cells in parasite infections. In: Galli SJ, Austen KF. *Mast cell and basophil differentiation and function in health and disease.* New York: Raven Press;1989:205-16.
27. Galli SJ, Tsai M, Wershil BK. The c-*kit* receptor, stem cell factor, and mast cells. What each is teaching us about the others. *Am J Pathol* 1993;142:965-74.
28. Wershil BK, Tsai M, Geissler EN, Zsebo KM, Galli SJ. The rat c-*kit* ligand, stem cell factor, induces c-*kit* receptor-dependent mouse mast cell activation *in vivo*. Evidence that signaling through the c-*kit* receptor can induce expression of cellular function. *J Exp Med* 1992;175:245-55.
29. Haig DM, Huntley JF, MacKellar A *et al*. Effects of stem cell factor (Kit-Ligand) and interleukin-3 on the growth and serine protease expression of rat bone-marrow-derived or serosal mast cells. *Blood* 1994;83:72-83.
30. Grencis RK, Else KJ, Huntley JF, Nishikawa SI. The *in vivo* role of stem cell factor (c-*kit* ligand) on mastocytosis and host protective immunity to the intestinal nematode *Trichinella spiralis* in mice. *Parasite Immunol* 1993;15:55-9.
31. Leftwich JA, Westin EH, Huff TF. Expression of c-*kit* by mesenteric lymph node cells from Nippostrongylus-brasiliensis-infected mice and by mast cell colonies developing from these cells in response to 3T3 fibroblast-conditioned medium. *J Immunol* 1992;148:2894-8.
32. Tsai M, Shih L-S, Newlands GFJ *et al*. The rat c-*kit* ligand, stem cell factor, induces development of connective tissue-type and mucosal mast cells *in vivo*. Analysis by anatomical distribution, histochemistry, and protease phenotype. *J Exp Med* 1991;174:125-31.
33. Cooper ES, Spencer J, Whyte-Alleng CA *et al*. Immediate hypersensitivity in colon of children with chronic *Trichuris trichiura* dysentery. *Lancet* 1991;338:1104-7.

19

Suppressive effect of interleukin-2 on the histamine release from rat peritoneal mast cells in connection with lipocortin-1 formation

Kenji Tasaka and Mitsunobu Mio

Department. of Pharmacology, Faculty of Pharmaceutical Sciences, Okayama University, Okayama, Japan.

It has been shown that the development and growth of mast cells are dependent upon interleukin-3 (IL-3), IL-4 and IL-9 (1 - 3). On the other hand, it has also been reported that histamine release from mast cells or basophils is induced or potentiated by several cytokines, such as IL-1α, IL-1β, IL-2, IL-3, IL-4 and granulocyte-macrophage colony stimulating factor (GM-CSF) (4 - 6). However, most of these studies are carried out using basophils and mucosal type mast cells, but not connective tissue type mast cells. Levi-Schaffer et al. (5) reported that IL-2, IL-3 and IL-4 enhanced both the basal and anti-IgE-mediated histamine release from rat and murine peritoneal mast cells co-cultured with fibroblasts. Moreover, there has been no demonstration about the inhibitory effect of cytokine on the histamine release from mast cells.

On the other hand, it has been shown that corticosteroids are effective in inhibiting histamine release from mast cells (7, 8). Since it has been indicated that corticosteroids induce the production of lipocortins and that lipocortins inhibited histamine release from mast cells, it was suggested that the inhibitory effect of corticosteroids on the histamine release can be ascribed to the lipocortin formation in mast cells (7, 8). However, there has been no report concerning the effect of cytokines on the lipocortin production in mast cells. In the present study, the inhibitory mechanism of the action of cytokines, especially IL-2 on the histamine release from rat peritoneal mast cells, was investigated.

INHIBITORY EFFECTS OF IL-2 ON HISTAMINE RELEASE FROM RAT PERITONEAL MAST CELLS INDUCED BY COMPOUND 48/80 AND CONCANAVALIN A

When rat peritoneal mast cells were pretreated with several cytokines at 37°C for 8 hr, the histamine release induced by compound 48/80 was significantly inhibited only by IL-2, but other cytokines such as IL-1α, IL-3, IL-4 and IL-5 were not effective at all, as shown in Table 1. As indicated in Fig. 1a, IL-2 inhibited the histamine release in a concentration-dependent manner: significant inhibition was observed at concentrations higher than 50 units/ml. When rat mast cells were incubated with IL-2 for various periods of time, the inhibitory effect of IL-2 on the histamine release became significant at 6 hr and the maximal inhibition achieved at 8 hr (Fig. 1b). When rat mast cells were exposed to concanavalin A at 30 and 100 μg/ml, histamine release was elicited at 27.2 ± 1.46 and 50.48 ± 2.39 %, respectively (n = 5). When the cells were pretreated with 100 units/ml of IL-2 for 8 hr prior to concanavalin A stimulation, the histamine release induced by 30 and 100 μg/ml of concanavalin A was significantly reduced to 20.65 ± 1.04 ($p < 0.05$) and 26.79 ± 1.72 % ($p < 0.01$), respectively.

Table 1. Effects of various cytokines on the histamine release from rat peritoneal mast cells induced by compound 48/80

cytokines	% histamine release	
	compound 48/80 alone	+ cytokines
IL-1α (100 units/ml)	41.7 ± 1.7	42.5 ± 1.3
IL-2 (100 units/ml)	46.5 ± 3.9	19.0 ± 1.5 **
IL-3 (50 units/ml)	50.9 ± 5.2	52.4 ± 2.7
IL-4 (50 ng/ml)	43.7 ± 2.6	42.3 ± 2.2
IL-5 (500 units/ml)	41.3 ± 4.2	45.4 ± 1.9

Rat peritoneal mast cells were incubated with various cytokines in DMEM at 37°C for 8 hr. Subsequently, the cells were incubated with compound 48/80 (1 μg/ml) for 10 min. ** indicates statistical significance at $p < 0.01$.

EFFECT OF IL-2 ON ^{45}CA UPTAKE AND IP$_3$ PRODUCTION IN RAT PERITONEAL MAST CELLS

It is known that an increase in intracellular Ca^{2+} concentration is a prerequisite for histamine release from mast cells (14). In addition, it has been shown that the IP$_3$ content in mast cells increases when the cells are treated with histamine-releasing stimuli and that this induces Ca^{2+} release from intracellular Ca store (15). In order to study the inhibitory mechanism of IL-2 on the histamine release from mast cells, the effects of IL-2 on the ^{45}Ca uptake and IP$_3$ production in rat peritoneal mast cells were investigated.

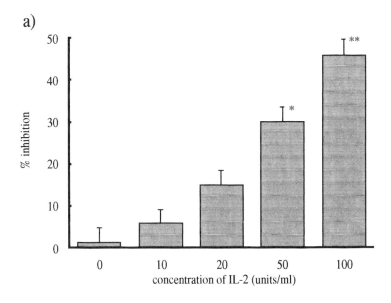

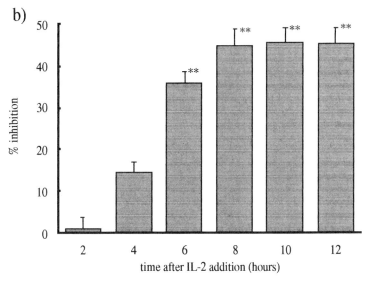

Fig. 1. Inhibitory effect of IL-2 on the histamine release from rat peritoneal mast cells induced by compound 48/80 (1 μg/ml). (a) Concentration-response relationship. Rat peritoneal mast cells were incubated with various concentrations of IL-2 for 8 hr. (b) Time course. Rat peritoneal mast cells were incubated with 100 units/ml of IL-2 for various periods of time. Each point represents the mean±S.E.M. of 5 separate experiments. * and ** indicate statistical significance in comparison with the control at $p<0.05$ and $p<0.01$, respectively.

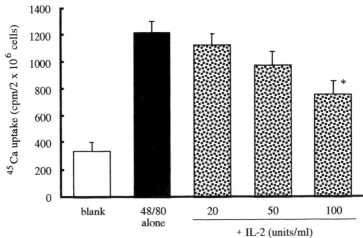

Fig. 2. Inhibitory effect of IL-2 on ^{45}Ca uptake into rat peritoneal mast cells elicited by compound 48/80 (1 µg/ml). Rat mast cells were incubated with various concentrations of IL-2 at 37°C for 8 hr. Thereafter, the cells were stimulated with compound 48/80 (1 µg/ml). Each point represents the mean±S.E.M. of 5 separate experiments.
* indicates statistical significance in comparison with control at $p<0.05$.

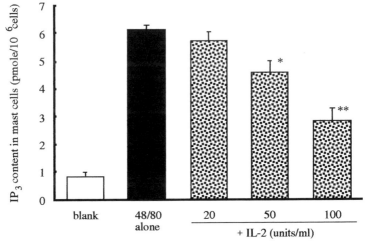

Fig. 3. Influence of IL-2 on the increase in IP$_3$ contents in rat peritoneal mast cells elicited by compound 48/80 (1 µg/ml). Rat peritoneal mast cells were incubated with various concentrations of IL-2 at 37°C for 8 hr. Thereafter, the cells were stimulated with 1 µg/ml of compound 48/80 for 5 sec. Each column represents the mean±S.E.M. of 4 separate experiments. * and ** indicate statistical significance in comparison with control at $p<0.05$ and $p<0.01$, respectively.

When rat peritoneal mast cells were stimulated by compound 48/80 at 1 μg/ml, ^{45}Ca uptake into mast cells increased to 3.5 times of the spontaneous uptake. However, ^{45}Ca uptake elicited by compound 48/80 was inhibited by IL-2 pretreatment at concentrations higher than 20 units/ml. Significant inhibition was observed at 100 units/ml (Fig. 2). IP$_3$ content in rat mast cells increased more than 9 times of the control value 5 sec after addition of compound 48/80 (2 μg/ml), as reported previously (12). IL-2 inhibited the compound 48/80-induced increase in IP$_3$ content in mast cells in a concentration-dependent manner (Fig. 3). Significant inhibition was observed at concentrations higher than 50 units/ml.

EFFECT OF IL-2 ON cAMP-PROTEIN KINASE A SYSTEM IN RAT MAST CELLS

It is known that an increase of cAMP content in mast cells and resulting activation of protein kinase A are effective in inhibiting the histamine release from mast cells (16). In order to study whether or not IL-2 induced inhibition is related with an activation of cAMP-protein kinase A system, the effect of protein kinase A inhibitors was studied on the histamine release inhibition elicited by IL-2 pretreatment. The inhibitory effect of IL-2 (100 units/ml) on the compound 48/80-induced histamine release was totally unaffected by protein kinase A inhibitors, such as H-8 (0.1 - 30 μg/ml) and KT-5720 (0.01 - 10 μg/ml). Furthermore, the cAMP contents in rat mast cells (2.812 ± 0.287 pmol/10^6 cells, n = 4) were not affected by 100 units/ml of IL-2 (2.743 ± 0.184 pmol/10^6 cells, n = 4). When rat mast cells were stimulated with 1 μg/ml of compound 48/80, the cAMP contents were decreased in both non-treated control (1.457 ± 0.262 pmol/10^6 cells, n = 4) and IL-2 treated cells (1.76 ± 0.368 pmol/10^6 cells, n = 4).

[^{3}H]-LEUCINE UPTAKE INTO RAT MAST CELLS ELICITED BY IL-2

Since IL-2 requires a long incubation period (more than 4 hr) to be effective in inhibiting the histamine release from mast cells, it was supposed that some proteins may be produced during the incubation period. To make this point clear, [^{3}H]-leucine uptake into rat mast cells was tested along with IL-2 incubation. When rat peritoneal mast cells were incubated with [^{3}H]-leucine in the absence of IL-2, the [^{3}H]-leucine content gradually increased as time elapsed, indicating that a spontaneous protein synthesis takes place (Fig. 4). On the other hand, when rat peritoneal mast cells were incubated with [^{3}H]-leucine in the presence of 100 units/ml of IL-2, significant enhancement of [^{3}H]-leucine uptake was observed. The most remarkable increase in the rate of [^{3}H]-leucine uptake takes place from 0.5 to 2 hr after addition of IL-2.

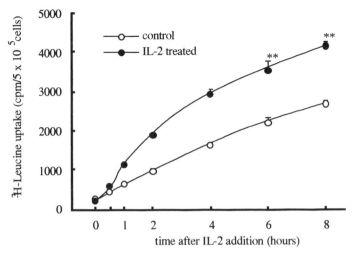

Fig. 4. Sequential changes in [3H]-leucine uptake into rat peritoneal mast cells induced by 100 units/ml of IL-2. Rat mast cells were incubated with [3H]-leucine both in the presence and in the absence of 100 units/ml of IL-2 for various periods of time. Thereafter, the cells were disrupted with TCA and the radioactivity incorporated into the mast cell proteins was determined. Each point represents the mean±S.E.M. of 4 separate experiments. * and ** indicate statistical significance in comparison with control at $p<0.05$ and $p<0.01$, respectively.

DETERMINATION OF THE PROTEIN SYNTHESIS ELICITED BY IL-2

In order to examine the protein(s) synthesized in mast cells after addition of IL-2, rat peritoneal mast cells were incubated with 100 units/ml of IL-2 in the presence of [3H]-leucine, and autoradiography of SDS-PAGE gel of mast cell proteins was carried out. The most obvious incorporation was noticed in the protein band having the molecular weight of about 35 kDa. When the SDS-polyacrylamide gel was cut into small pieces after electrophoresis and the radioactivity of each piece was measured, it was indicated that [3H]-leucine uptake into the proteins having molecular weights of 30 - 35 kDa increased more than 4 times the control level (Fig. 5).

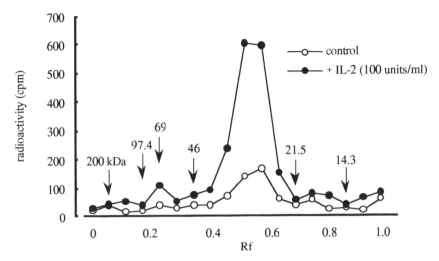

Fig. 5. Quantitative analysis of [3H]-leucine uptake in mast cell proteins induced by IL-2. Rat peritoneal mast cells were incubated with 100 units/ml of IL-2 in the presence of 1 μCi/ml of [3H]-leucine at 37°C for 8 hr. After SDS-PAGE of mast cell proteins, the gel was cut into small pieces and the radioactivity was measured.

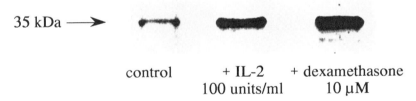

Fig. 6. Western blotting analysis of mast cell proteins using anti-lipocortin-I antibody. Rat peritoneal mast cells were incubated with 100 units/ml of IL-2 or 10 μM of dexamethasone at 37°C for 8 hr. Thereafter, Western blotting analysis of mast cell proteins was carried out using an anti-lipocortin-I antibody.

WESTERN BLOTTING ANALYSIS OF LIPOCORTIN-I INDUCED BY IL-2

The newly synthesized protein in mast cells produced after IL-2 exposure seems to possess characteristics similar to lipocortin, since 1) IL-2 is able to inhibit IP_3 production in mast cells and 2) molecular weight of newly synthesized protein is about 30 - 35 kDa (7, 17). To make this point clear, Western blotting analysis of mast cell proteins was carried out using an anti-lipocortin-I antibody. As shown in Fig. 6, in the resting state, rat peritoneal mast cells contained only a small amount of lipocortin-I with the molecular weight of 35 kDa. When the cells were incubated with 100 units/ml of IL-2 for 8 hr, the content of lipocortin-I markedly increased, indicating that IL-2 caused a significant increase of lipocortin-I formation. Similar increase in lipocortin-I production was also observed when mast cells were incubated with 10 μM of dexamethasone at 37°C for 8 hr.

PARTICIPATION OF TYROSINE KINASE IN HISTAMINE RELEASE INHIBITION DUE TO IL-2

It has been shown that the IL-2 receptor possesses tyrosine kinase activity (18). To make clear whether or not tyrosine kinase plays an important role in inhibiting histamine release, the influence of the tyrosine kinase inhibitor, genistein, was investigated on the histamine release inhibition caused by IL-2. As shown in Fig. 7, the inhibitory effect of IL-2 on the histamine release elicited by compound 48/80 was dose-dependently abrogated by genistein. When rat mast cells that had not been exposed to IL-2 were pretreated with genistein at concentrations ranging from 0.3 to 30 μM, the histamine release from mast cells induced by compound 48/80 (1 μg/ml) was not affected (data not shown). When the phosphorylation profile of mast cell protein was determined, it was indicated that the phosphorylation of 18 kDa protein was markedly increased 1 min after IL-2 addition (Fig. 8). When the radioactivity of phosphorylated protein bands was measured, the phosphorylation of 18 kDa protein reached the maximum, 1 min after IL-2 addition and decreased to the control level at 15 min (Table 2). Such sequential changes were consistently observed.

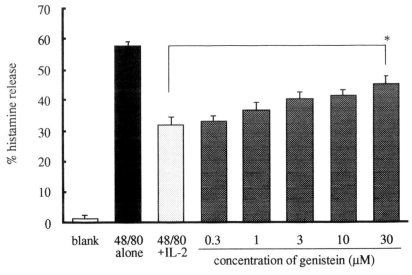

Fig. 7. Effect of the tyrosine kinase inhibitor, genistein, on the histamine release inhibition due to IL-2. Rat peritoneal mast cells were exposed to genistein 30 min prior to IL-2 addition. Thereafter, the mast cells were incubated with 100 units/ml of IL-2 at 37°C for 8 hr. After that, the cells were stimulated by 1 mg/ml of compound 48/80 at 37°C for 10 min. Each column represents the mean±S.E.M. of 5 separate experiments. * indicates statistical significance in comparison with control at $p<0.05$.

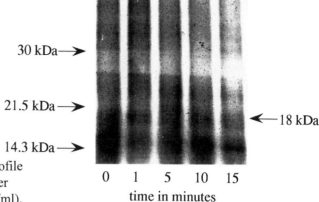

Fig. 8. Phosphorylation profile of rat mast cell proteins after addition of IL-2 (100 units/ml).

Table 2. Sequential changes in the phosphorylation of 18 kDa protein in rat mast cells induced by IL-2 (100 units/ml).

Time in minutes	radioactivity (% of control)
0	100.0
1	137.9
5	116.1
10	108.6
15	103.2

Rat peritoneal mast cells were incubated with 100 units/ml of IL-2 at 37°C for various periods of time in the presence of 500 µCi/ml of [^{32}P]-orthophosphate. Thereafter, SDS-PAGE of mast cell proteins was carried out. Radioactivity of 18 kDa protein band was determined by means of Image Analyzer BAS-2000.

DISCUSSION

It has been suggested that there exists IL-2 receptor on human mast cells and basophils (19). Levi-Schaffer et al. (5) reported that IL-2 enhanced histamine release from rat peritoneal mast cells co-cultured with fibroblasts. However, there has been no report about the inhibitory effect of cytokines on the histamine release from mast cells. In the present study, it became apparent that IL-2 inhibited histamine release from rat peritoneal mast cells. This result seems to be contradictory to that reported by Levi-Schaffer et al. (5). Although Levi-Schaffer et al. cultured rat peritoneal mast cells with fibroblasts, we examined the direct action of IL-2 on mast cells. Therefore, it was supposed that in the results of Levi-Schaffer et al. (5) the direct action of IL-2 may be modified by the indirect action of IL-2 which was elicited by co-culture with fibroblasts. In the present experiment, other cytokines did not affect histamine release from rat peritoneal mast cells induced by compound 48/80. It has been reported that IL-3 and IL-4 enhance histamine release from mucosal type mast cells and basophils (4, 6). Moreover, when bone marrow cells are co-cultured with IL-3, the cells differentiated to bone marrow-derived mast cells, which can be classified as mucosal type mast cells (1). Therefore, it was assumed that connective tissue type mast cells, such as rat peritoneal mast cells, may not have the receptors for IL-3 and IL-4, and the differentiation pathway of connective tissue mast cells may be different from that of mucosal type mast cells.

As shown in the present results, IL-2 inhibited ^{45}Ca uptake (Fig. 2) and IP$_3$ production (Fig. 3) in rat peritoneal mast cells stimulated by compound 48/80 without affecting intracellular cAMP levels. Protein kinase A inhibitors did not affect the histamine release inhibition due to IL-2. These results clearly indicate that IL-2 prevented the histamine release from mast cells by inhibiting the increase in intracellular Ca^{2+} concentrations without affecting cAMP-protein kinase A system. Although it has been reported that an activation of cAMP-protein kinase A system is effective in inhibiting an increase in intracellular Ca^{2+} concentration (14, 16), the present results indicate that something other than cAMP may be involved in the histamine release inhibition due to IL-2. Since an increase in intracellular Ca^{2+} concentration is a prerequisite for the histamine release from mast cells, the histamine release inhibition due to IL-2 can be ascribed to the inhibitions of IP$_3$ production and Ca uptake. Since it requires several hours to exhibit the inhibitory activity of IL-2 on the histamine release from mast cells and IL-2 enhanced [^{3}H]-leucine uptake into mast cells, it was supposed that the histamine release inhibition may be associated with protein synthesis in mast cells. Actually, IL-2 stimulated the synthesis of protein, which has a molecular weight of about 35 kDa and was determined as lipocortin-I based on Western blotting analysis. Lipocortin-II was not detected in mast cell proteins (data not shown). In accordance with this, it has been shown that glucocorticoids are effective in inhibiting histamine release from mast cells and that this inhibition may be intimately related to the production of lipocortins (7, 8). The production of lipocortin-I in the presence of dexamethasone was also confirmed in the present study as shown in Fig. 6. In both cases, lipocortin production became evident at 4 hr after additions, and reached the maximum at 8 hr. The time course of the protein synthesis almost corresponds with that of histamine release inhibition induced by IL-2, suggesting that lipocortin-I production may be intimately related to histamine release inhibition.

It is well known that lipocortin inhibits not only phospholipase A$_2$ but also phospholipase C, especially polyphosphoinositide-specific phospholipase C, and this subsequently results in an inhibition of IP$_3$ production (17, 20). As shown in Fig. 3, IL-2 inhibited IP$_3$ production, which is an essential trigger for Ca^{2+} release from intracellular Ca store, and, also, the subsequent histamine release from mast cells (15). Therefore, it was assumed that the histamine release inhibition due to IL-2 could be ascribed to the production of lipocortin-I and the resulting inhibitions of phospholipase C and IP$_3$ production. In addition, since it has been reported that inositol 1,3,4,5-tetrakisphosphate (IP$_4$), which is a phosphorylated product of IP$_3$, stimulates the opening of the Ca channel (21), the inhibitory effect of IL-2 on ^{45}Ca uptake may also be ascribed to the inhibition of phospholipase C and resulting inhibition of IP$_3$ and IP$_4$ productions.

It has been indicated that the IL-2 receptor possesses tyrosine kinase activity (18). As shown in Fig. 7, the histamine release inhibition due to IL-2 was reversed by genistein, an inhibitor of tyrosine kinase, in a dose-dependent manner. A phosphorylation of 18 kDa protein in mast cells was also elicited by IL-2 treatment. These results may indicate that tyrosine kinase probably participates in histamine release inhibition due to IL-2. Although the function of 18 kDa protein in mast cells is not clear at the present time, it may be reasonable to assume that IL-

2 may increase tyrosine kinase activity of the IL-2 receptor so as to phosphorylate 18 kDa protein and, as a consequence of this, an increase of the lipocortin-I production may take place. The time course of the phosphorylation and dephosphorylation of the 18 kDa protein is much more rapid than that seen in lipocortin-I formation. This may suggest that the phosphorylation of 18 kDa protein may be involved in the signal transduction from the IL-2 receptor to the nucleus in the process leading to lipocortin-I production. The newly synthesized lipocortin-I seems to be effective in inhibiting the phospholipase C activity and this may be effective in inhibiting histamine release from mast cells.

In the case of the glucocorticoid-induced lipocortin production, it has been indicated that the intracellular glucocorticoid receptor participates in the gene expression of lipocortins (22). On the other hand, it has been shown that IL-2 induces several gene expressions and/or differentiations in various types of cells (23 - 26). At present time, it is still uncertain whether or not IL-2 induces the gene expression of lipocortins or whether the pathway of IL-2-induced lipocortin-I production shares the same way of glucocorticoid. However, the fact that both glucocorticoid and IL-2 induced lipocortin-I production in mast cells may suggest that a similar mechanism may exist between the IL-2-induced and glucocorticoid-induced gene expression of lipocortins.

From the present study, it was concluded that IL-2 induces lipocortin-I production in rat peritoneal mast cells. The newly synthesized lipocortin-I inhibits an activation of phospholipase C and, as a consequence of this, an increase in intracellular Ca^{2+} concentration may be blocked. This may further explain the histamine release inhibition from mast cells.

REFERENCES

1. Ihle JN, Keller J, Oroszlan S et al. Biologic properties of homogeneous interleukin 3. I. Demonstration of WEHI-3 growth factor activity, mast cell growth activity, P cell-stimulating factor activity, colony-stimulating factor activity, and histamine-producing cell-stimulating factor activity. *J Immunol* 1983; 131: 282-287
2. Schmitt E, Fassbender B, Beyreuther K, Spaeth E, Schwarzkopf R and Ruede E. Characterization of a T cell-derived lymphokine that acts synergistically with IL-3 on the growth of murine mast cells and is identical with IL-4. *Immunobiology* 1987; 174: 406-419
3. Hultner L, Druez C, Moeller J et al. Mast cell growth-enhancing activity (MEA) is structurally related and functionally identical to the novel mouse T cell growth factor P40/TCGFIII (interleukin 9). *Eur J Immunol* 1990; 20: 1413-1416
4. Massey AW, Randall CT, Kagey JA et al. Recombinant human IL-1α and -1β potentiate IgE-mediated histamine release from human basophils. *J Immunol* 1989; 143: 1875-1880
5. Levi-Schaffer F, Segal V and Shalit M. Effect of interleukins on connective tissue type mast cells co-cultured with fibroblasts. *Immunology* 1991; 72: 174-180
6. Haak-Frendscho M, Arai N. and Arai K. Human recombinant granulocyte-macrophage colony-stimulating factor and interleukin-3 cause basophil histamine release. *J Clin Invest* 1988; 82: 17-20

7. White MV, Igarashi Y, Lundgren JD, Shelhamer J and Kaliner M. Hydrocortisone inhibits rat basophilic leukemia cell mediator release induced by neutrophil-derived histamine releasing activity as well as by anti-IgE. *J Immunol* 1991; 147: 667-673
8. Sautebin L, Carnuccio R, Ialenti A and Di Rosa M. Lipocortin and vasocortin: Two species of anti-inflammetory proteins mimicking the effects of glucocorticoids. *Pharmacol Res* 1992; 25: 1-12
9. Nemeth A and Röhlich P. Rapid separation of rat peritoneal mast cells with Percoll. *Eur J Cell Biol* 1980; 20: 272-275
10. Siraganian RP. An automated continuous-flow system for the extraction and fluorometric analysis of histamine. *Anal Biochem* 1974; 57: 383-394
11. Spataro AC and Bosmann HB. Mechanism of action of disodium cromoglycate – mast cell calcium ion influx after a histamine-releasing stimulus. *Biochem Pharmacol* 1976; 25: 505-510
12. Tasaka K, Mio M, Fujisawa K and Aoki I. Role of microtubules on Ca^{2+} release from the endoplasmic reticulum and associated histamine release from rat peritoneal mast cells. *Biochem Pharmacol* 1991; 41: 1031-1037
13. Laemmli UK. Cleavage of structural proteins during the assembly of the head of bacteriophage T4. *Nature* 1970; 227: 680-685
14. Tasaka K, Mio M and Okamoto M. Intracellular calcium ion release induced by histamine releasers and its inhibition by some antiallergic drugs. *Ann Allergy* 1986; 56: 464-469
15. Tasaka K, Sugimoto Y and Mio M. Sequential analysis of histamine release and intracellular Ca^{2+} release from murine mast cells. *Int Arch Allergy Appl Immunol* 1990; 91: 211-213
16. Izushi K, Shirasaka T, Chokki M and Tasaka K. Phosphorylation of smg p21B in rat peritoneal mast cells in association with histamine release inhibition by dibutyryl-cAMP. *FEBS Lett* 1992; 314: 241-245
17. Machoczek K, Fischer M and Soling H-D. Lipocortin I and lipocortin II inhibit phosphoinositide- and polyphosphoinositide-specific phospholipase C. The effect results from interaction with the substrates. FEBS Lett 1989; 251: 207-212
18. Waldmann AT. The interleukin-2 receptor. *J Biol Chem* 1991; 286: 2681-2684
19. Maggiano M, Cotta F, Castellino F et al. Interleukin-2 receptor expression in human mast cells and basophils. *Int Arch Allergy Appl Immunol* 1990; 91: 8-14
20. Weber G and Ferber E. Selective inhibition of phospholipase A_2 by different lipocortins. *J Biol Chem* 1990; 371: 725-731
21. Pittet D, Lew DP, Mayr GW, Monod A and Schlegel W. Chemoattractant receptor promotion of Ca^{2+} influx across the plasma membrane of HL-60 cells. A role for cytosolic free calcium elevations and inositol 1,3,4,5-tetrakisphosphate production. J Biol Chem 1989; 264: 7251-7261
22. Flower RJ. Lipocortin and the mechanism of action of the glucocorticoids. *Br J Pharmacol* 1988; 94: 987-1015
23. Emilie D, Karray S, Merle-Beral H, Debre P and Galanaud P. Induction of differentiation in human leukemic B cells by interleukin 2 alone: Differential effect on the expression of μ and J chain genes. *Eur J Immunol* 1988; 18: 1479-1483
24. Espinoza-Delgado I, Longo DL, Gusella GL and Varesio L. IL-2 enhances c-fms expression in human monocytes. *J Immunol* 1990; 145: 1137-1143

25. Smyth MJ, Ortaldo JR, Bere W, Yagita H, Okumura K and Young HA. IL-2 and IL-6 synergize to augment the pore-forming protein gene expression and cytotoxic potential of human peripheral blood T cells. *J Immunol* 1990; 145: 1159-1166
26. Sabath DE, Broome HE and Prystowsky MB. Glyceraldehyde-3-phosphate dehydrogenase mRNA is a major interleukin 2-induced transcript in a cloned T-helper lymphocyte. *Gene* 1990; 91: 185-191

20

Fibronectin Receptor (FNR) Integrins On Mast Cells Are Involved In Cellular Activation

Chisei Ra, Masahiko Yasuda, Zaisun Kim, *Hirohisa Saito, **Tatsutoshi Nakahata, Hideo Yagita, and Ko Okumura

Department of Immunology, Juntendo University, School of Medicine, Tokyo 113, Japan
*Division of Allergy, National Children's Medical Research Center, Tokyo 154, Japan
**Department of Clinical Oncology, The Institute of Medical Science, The University of Tokyo 108, Japan

Adhesion molecules of immunogloblin and integrin superfamilies are implicated not only in leukocyte interaction and migration but also in cellular activation as described well with T cells (16). Aggregation of the high affinity receptor for IgE (FcεRI) triggers mast cell activation for degranulation and cytokine secretion, and this signal transduction through FcεRI is considered very similar to that through TCR/CD3 in T cells, both of which involve PTK activation, tyrosine phospholyration of signal transducing subnits, phospholipase C activation, and subsequent Ca^{2+} mobilization and PKC activation (9, 12, 14). Of signal transducing subnits of these receptors, CD3-ζ and FcεRIγ chains are structurally similar molecules belonging to a same family and have a common motif containing two tyrosine residues for signal transduction (2, 11, 13). It is suggested that tyrosine residues in this motif are phospholyrated by src family kinases associating with CD3-ζ or FcεRIγ chain upon receptor engagement (4, 10). Further ZAP70 and Syk kinase associate with CD3-ζ and FcεRIγ chain respectively on receptor cross-linking (1, 3). These common events observed in an early signal transduction pathway of TCR/CD3 and FcεRI give a good rationale to test a hypothesis that signals *via* adhesion molecules on mast cells also modulate mast cell activation as those on T cells do in T cell activation.

Mast cells in peripheral tissues are actually surrounded by other cells such as fibroblasts, mucosal cells and extracellular matrix (ECM) proteins (collagen, fibronectin, laminin etc.), suggesting that mast cells adhere to these cells and ECM receiving some signals through adhesion molecules expressed on the cell surface (6). Therefore to investigate the process of allergic inflammation *in vivo,* we need to evaluate the effect of integrin-mediated signals on mast cell activation.

I. MAST CELLS ADHERE SPECIFICALLY TO ECM PROTEINS

We first examined the adherence of mast cells to various ECM proteins coated on the plastic plate. A rat mucosal mast cell line, RBL-2H3, bound to fibronectin (FN), vitronectin (VN) and fibrinogen (FB) without any stimuli, but not to laminin (LM) or collagen (CL). Freshly prepared rat peritoneal mast cell (RPMC) exhibited a similar binding specificity to ECM proteins but only after PMA stimulation like the case of resting T cells where the $\beta 1$ and $\beta 2$ integrin-mediated adhesion was activated by PMA (7, 8, 15). A mouse connective tissue type mast cell line, PT18, bound only to FN and neither to VN, FB, LM, nor to CL. Mouse bone marrow derived mast cells cultured with IL-3 (CMC) bound to FN, VN, and to a lesser extent LM when stimulated with PMA. A human basophilic leukemia cell line, KU812, with PMA-stimulation, bound to FN and FB (to a lesser extent) but neither to VN, LM, nor to CL.

The FN-adherence of RBL-2H3 and RPMC was inhibited by both conecting segment (CS)-1 peptide of FN that blocks very late activation antigen (VLA)-4 interaction with FN and RGD peptide of FN that blocks VLA-5 and VN receptor (VNR) interaction with FN. In the case of PT18, FN-adherence was strongly inhibited by RGD peptide and weakly inhibited by CS-1 peptide (Table 1).

Table 1. Adhesion of mast cells to ECM proteins

ECM	Rodent cells				Human cells
	RBL-2H3	RPMC	PT18	CMC	KU812
FN	###	###	###	###	##
VN	##	###	—	##	—
FB	+	###	—	±	+
CL	—	—	—	±	—
LM	±	+	—	+	—

Abbreviations : ECM, extracellular matrix protein; RBL-2H3, rat basophilic-leukemia cell line; RPMC, rat peritoneal mast cell; PT18, mouse mast-cell line; CMC, cultured mast cell derived from bone marrow; KU812, human basophilic leukemia cell line; FN, fibronectin; VN, vitronectin; FB, fibrinogen; CL, collagen; LM, laminin; ### ~ ± , cells adhere to the ligands; —, almost no cells adhere to the ligands.

II. EXPRESSION OF ADHESION MOLECULES ON MAST CELLS

We confirmed the expression of receptors on mast cells for ECM proteins by FACS analysis using anti-integrin monoclonal antibodies (mAbs). RBL-2H3 expressed VLA-α4, VLA-α5, and VNR(β3) on the cell surface and RPMC also expressed VLA-α4, VNR(β3) and to a lesser extent VLA-α5. Anti-VLA-α4, -VLA-α5 and -VNR(β3) mAbs inhibited FN-adherence of both RBL-2H3 and RPMC, and the strongest inhibitory effect was observed with the mixture of all these antibodies, which was consistent with the above mentioned inhibitory effects of RGD and CS-1 peptides.

Both PT-18 and CMC, primaly cultured mouse mast cells, expressed VLA-α4, VLA-α5 and VNR(β3) on their cell surface like rat mast cells. On CMC, the expression of VLA-α1, VLA-α2 and VLA-α6 as LM receptors was also observed. FN-adherence of PT18 was strongly inhibited by anti-VLA-α5 and -VNR(β3) mAbs as by RGD peptide, and weakly inhibited by anti-VLA-α4 mAb as by CS-1 peptide, indicating that VLA-5 is dominant as a FNR in PT18. With the mixture of these mAbs, the strongest inhibition was observed in PT18. FN-adherence of CMC was weakly blocked by each antibody (anti-VLA-α4, VLA-α5, VNR(β3)) and strongly blocked by the mixture of these antibodies. LM-adherence of CMC was weakly blocked by anti-VLA-α2, -VLA-α6 mAbs and strongly blocked by the mixture of these mAbs.

KU812 strongly expressed VLA-α4, α5 and to a lesser extent VLA-α2, α3 but not VLA-α1. Intercellular adhesion molecule (ICAM)-1, ICAM-2 were also strongly expressed on KU812. Human cultured mast cells (derived from stem cells) expressed VLA-α2, α3, α4, α5, and VNR but not VLA-α1, α6. Leukocyte adhesion molecule (LFA)-1, 2, 3, and ICAM 1, 2 were also expressed in human cultured mast cells (Table 2).

Table 2. Expression of adhesion molecules on mast cells

Integrin		Ligand	MouseCMC	HumanCMC
VLA-1	$\alpha_1\beta_1$	LM/CL	+	−
VLA-2	$\alpha_2\beta_1$	CL/LM	+	+
VLA-3	$\alpha_3\beta_1$	FN/LM/CL	N.D.	++
VLA-4	$\alpha_4\beta_1$	FN/VCAM-1	++	++
VLA-5	$\alpha_5\beta_1$	FN	++	+
VLA-6	$\alpha_6\beta_1$	LM	+	−
LFA-1	$\alpha_L\beta_2$	ICAM-1/ICAM-2	N.D.	+
Mac-1	$\alpha_M\beta_2$	C3bi/FB/ICAM-2	N.D.	±
P150,95	$\alpha_X\beta_2$	C3bi/FB	N.D.	N.D
VNR	$\alpha_V\beta_3$	VN/FB/FN/vWF	++	+
ICAM-1		LFA-1	N.D.	+
ICAM-2		LFA-1	N.D.	N.D.
LFA-3		LFA-2	N.D.	+

Abbreviations: VLA, very late activation antigen; LFA, leukocyte function-associated antigen; VNR, vitronectin receptor; ICAM, intercellular adhesion molecule; C, complement; vWF, von Willebrand factor; ++ ~ ±, molecules are present on the cell surface; –, molecules are absent on the cell surface; N.D., not done.

These results indicate that the adherence of mast cells to ECM proteins is mediated through the specific interaction of integrin molecules with ECM proteins.

III. THE ENGAGEMENT OF FNR INTEGRINS ENHANCES MAST CELL DEGRANULATION INDUCED BY FcεRI AGGREGATION.

Exocytosis of β-hexosaminidase as an indicator of degranulation from RBL-2H3 in response to FcεRI cross-linking with IgE and anti-IgE Ab was greatly enhanced on FN-coated plates specifically but not on CL- and BSA-coated plates. This enhancement of exocytosis on FN was blocked by RGD and CS-1 peptides or by anti-VLA-α4, -VLA-α5, and -VNR mAbs. Fab flagments of these Abs were as effective as whole Abs, indicating that this inhibition was due to the real blocking of the specific interaction between FN and integrins but not due to some modulatory signals *via* Fcγ receptors. Taken together, we confirmed FNR-mediated specific interaction with FN promotes mast cell activation induced by FcεRI-engagement.

The involvement of FNR in mast cell activation was investigated also *in vivo*. Local administration of the mixture of anti-VLA-α4, -VLA-α5, -VNR(β3) mAbs greatly reduced PCA (passive cutaneous anaphylaxis) reaction in rats, indicating that these antibodies dissociate mast cells from ECM or some other cells such as fibroblasts in local microenvironment. Especially when IgE and/or antigen level is low, enhancing signals for mast cell degranulation through integrins on the cell surface may play a significant role in allergic state.

IV. FNR INTEGRINS ON MAST CELLS ARE INVOLVED IN CYTOKINE SECRETION AND PRODUCTION.

It is now well established that mast cells are also important as a various cytokine source such as IL-3, IL-4, IL-5, TNF-α, GM-CSF and IFN-γ, inducing allergic inflammation (5).

Among these cytokines produced by mast cells, TNF-α and IL-4 are of great importance because these cytokines induce adhesion molecules on vascular endotherial cells in inflamed lesion to initiate the migration of inflammatory cells such as granulocytes and lymphocytes.

We first assessed the involvement of FNR integrins in TNF-α production and release by PT18 cells which contain a large amount of TNF-α. PT18 cells on FN with FcεRI engagement released TNF-α approximately 2.5 times more than PT18 cells on CL although almost no release of TNF-α was observed without FcεRI engagement. RGD peptide abolished this enhancement of TNF-α release on FN and CS-1 peptide inhibited this enhancing effect to a lesser extent, indicating that VLA-5 is dominant in enhancement of TNF-α release from PT18 cells on FN.

We next examined TNF-α gene expression in PT18 cells by the reverse transcriptase-polymerase chain reaction (RT-PCR) method. TNF-α gene was constitutively expressed without any stimuli. Cross-linking of FcεRI by IgE-anti-IgE Ab complexes greatly promoted TNF-α gene expression of PT18 cells to the same extent whether the cells were on FN or CL. Together with the results mentioned above, it was indicated that signals via FNR integrins on PT18 cells enhance TNF-α release but have no effect on TNF-α gene expression.

Cytokine gene expression in mouse CMC was also investigated by RT-PCR method. IL-3, IL-4 and GM-CSF genes were constitutively expressed without any stimuli. With FcεRI engagement, the expression of IL-3 and IL-4 gnes were greatly enhanced but GM-CSF gene-expression was not enhanced at all. Very interestingly, only IL-3 gene-expression was enhanced in CMC on FN, suggesting that CMC survival should be prolonged by autocrine/paracrine system of IL-3 secretion when the cells are on FN.

V. PROLONGED SURVIVAL OF MAST CELLS ON FIBRONECTIN

CMC was derived from mouse bone marrow and cultured with IL-3 for 4 weeks and then CMC survival rate was estimated on FN. Survival rate of CMC on FN was significantly higher than that on CL and BSA whether FcεRI were engaged or not. This prolongation of CMC survival on FN was blocked by anti-VLA-α4, -VLA-α5 and -VNR(β3)mAbs, indicating that this survival prolongation was really specific for FNR integrins' engagement on the cell surface. In addition, only anti-IL-3 mAb abolished mast cell

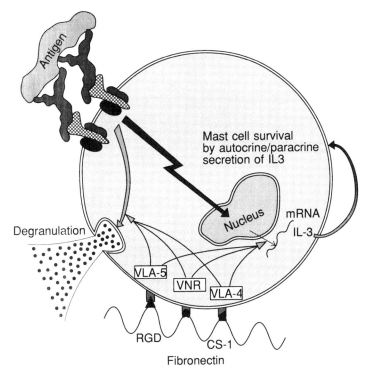

Figure. Involvement of FNR integrins in mast cell activation

Cross-linking of FcεRI by antigen-IgE complexes evokes intracellular signals leading to degranulation (➡) and on the other hand to cytokine-gene expression (⇀). Engagement of FNR integrins enhances the degranulation and IL-3 gene expression (⇒). Mast cell survival (mouse) on fibronectin is prolonged by autocrine/paracrine secretion of IL3.

survival on FN but not anti-IL-4, -GMCSF mAbs. Taken together, our findings suggest that IL-3 autocrine/paracrine system of mast cells on FN in peripheral tissues may play an important role for the mast cell survival.

CONCLUSIONS

We documented here the involvement of FNR integrins in mast cell activation. Although without cross-linking of FcεRI, mast cells on ECM proteins were not activated to degranulation, our findings suggest that FNR integrins' engagement on the cell surface enhances the sensitivity of mast cell activation and plays a significant role especially when antigen and IgE stay at low level in microenvironment. Another interesting finding is the survival prolongation of mouse mast cells on FN probably due to IL-3 autocrine/paracrine system (Figure).

Elucidation of molecular mechanisms underlying the phenomena we have found will enable us to further explore the allergic state and give us fine tools for the medical manipulation of allergy.

ACKNOWLEDGEMENTS

This work was supported in part by grants from Ministry of Health and Welfare (Japan) and SANDOZ Inc. (Japan). We would like to thank Mr. Matsuda H. for technical assistance and Miss. Ikeda Y. for secretarial help.

REFERENCES

1 Benhamou M, Ryba N J P, Kihara H, Nishika Protein-tyrosine kinase $p72^{syk}$ in high affinity IgE receptor signaling. *J Biol Chem* 1993:268:23318-23324.

2 Blank U, Ra C, Miller L, White K, Metzger H, Kinet J P. Complete structure and expression in transfected cells of high affinity IgE receptor. *Nature* 1989:337:187-189.

3 Chan A C, Iwashima M, Turck C W, Weiss A. ZAP-70: A 70 kd protein-tyrosine kinase that associates with the TCR ζ chain. *Cell* 1992:71:649-662.

4 Eiseman E, Bolen J. Engagement of the high-affinity IgE receptor activates src protein-related tyrosine kinases. *Nature* 1992:355:78-80.

5 Gordon J R, Burd P R, Galli S J. Mast cells as a source of multifunctional cytokines. *Immunol Today* 1990:11:458-464.

6 Hamawy M M, Mergenhagen S E, Siraganian R P. Adhesion molecules as regulators of mast-cell and basophil function. *Immunol Today* 1994:15:62-66.

7 Hamawy M H, Oliver C, Mergenhagen S E, Siraganian R P. Adherence of rat basophilic leukemia (RBL-2H3) cells to fibronectin-coated surfaces enhances secretion. *J Immunol* 1992:149:615-621.

8 Hamawy M H, Mergenhagen S E, Siraganian R P. Cell adherence to fibronectin and the aggregation of the high affinity immunoglobulin E receptor synergistically regulate tyrosine phosphorylation of 105-115-kDa proteins. *J Biol Chem* 1993:268:5227-5233.

9 Metzger H, Kinet J P, Blank U, Miller L, Ra C. The receptor with high affinity for IgE. In: Chadwick D, Evered D, Whelan J. *IgE, Mast cells and allergic response.* John Wiley and Sons; 1989:93-101.

10 Paolini R, Jouvin M H, Kinet J P. Phosphorylation and dephosphorylation of the high-affinity receptor for immunoglobulin E immediately after receptor engagement and disengagement. *Nature* 1991:353:855-858.

11 Ra C, Jouvin M H, Kinet J P. Complete structure of the mouse mast cell receptor for IgE (FcεRI) and surface expression of chimeric receptors (rat-mouse-human) on transfected cells. *J Biol Chem* 1989:264:15323-15327.

12 Ravetch J V, Kinet J P. Fc receptors. In: Paul W E, Fathman C G, Metzger H. *Annu Rev Immunol*, vol 9. Palo Alto: Annual Reviews Inc; 1991:457-492.

13 Reth M. Antigen receptor tail clue. *Nature* 1989:338:383-384.

14 Siraganian R P. Mechanism of IgE-mediated hypersensitivity. In: Middleton E Jr, Reed C E, Ellis E F, Adkinson N F Jr, Yunginger J W, Busse W W. *Allergy-Principles and practice*. 4th ed. Mosby; 1993:105-134.

15 Thompson H L, Burbelo P D, Yamada Y, Kleinman H K, Metcalfe D D. Mast cells chemotax to laminin with enhancement after IgE-mediated activation. *J Immunol* 1989:143:4188-4192.

16 Yagita H, Okumura K. The role of adhesion molecules in triggering effector mechanisms. In: Nariuchi H, Okada H, Okumura K, Takatsuki K, Yodoi J. *Molecular basis of immune responses*. Tokyo: Academic Press; 1993:59-70.

21
Tyrosine Kinases and Their Substrates in the FcεRI Signaling Pathway

Toshiaki Kawakami, Yuko Kawakami, Libo Yao, Hiromi Fukamachi, Sayako Matsuoka and Toru Miura.

Division of Immunobiology, La Jolla Institute for Allergy and Immunology, La Jolla, California 92037

Cross-linking of the high-affinity receptor for IgE (FcεRI) on mast cells and basophils induces a variety of biochemical events leading to the release of vasoactive amines, and the synthesis and secretion of several cytokines and lipid mediators. Among early changes are an increased production of inositol phosphates and diacylglycerol (DAG), and Ca^{2+} mobilization. Another consequence of these biochemical changes is cell proliferation, promoted by cytokines secreted by activated mast cells themselves and other types of cells. Recent studies have demonstrated tyrosine phosphorylation of a 72-kDa protein and other proteins as the earliest signaling event following FcεRI cross-linking (1). Our previous study with several tyrosine kinase inhibitors with different inhibitory modes has established an essential role of protein-tyrosine kinase(s) (PTK) in mast cell activation through FcεRI (2).

FcεRI is a tetrameric complex composed of an IgE-binding α subunit, a β subunit and a homodimer of disulfide-linked γ subunits. Fc receptors, including FcεRI, together with the T cell antigen receptor (TCR) and the B cell antigen receptor constitute a distinct class of cell surface receptors. They have ligand-binding subunit(s), all belonging to the immunoglobulin superfamily. Their signaling subunits, including the β and γ subunits of FcεRI, share a short cytoplasmic amino acid motif (D/E-X_7-D/E-X_2-Y-X_2-L/I-X_7-Y-X_2-L/I), termed either the tyrosine activation motif (TAM), antigen receptor homology 1 (ARH1) or antigen receptor activation motif (ARAM), which is of critical importance for cell activation. Unlike growth factor receptors such as the epidermal growth factor receptor, these receptors lack intrinsic tyrosine kinase activity. However, their cross-linking leads to rapid tyrosine phosphorylation of cellular proteins, and inhibition of tyrosine phosphorylation blocks cell activation.

At least two subfamilies of cytoplasmic PTKs serve as the candidate enzymes that are coupled to FcεRI, and phosphorylate key substrates that participate in the

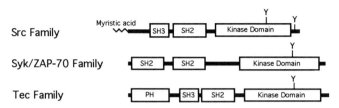

FIG. 1. Three PTK families implicated in the FcεRI signaling.

signaling pathway. Structural features of these PTKs are shown in Fig. 1. Src family PTKs, Lyn and c-Yes, were shown to be associated with FcεRI and activated upon cross-linking of the receptor (3). c-Src was also activated by FcεRI cross-linking. A member of another subfamily PTK, termed p72^{syk} (or PTK72), was also shown to be associated with FcεRI and its catalytic activity was activated upon receptor cross-linking (4). Later, the association of p72^{syk} with FcεRI γ subunit was shown to be dependent on FcεRI cross-linking (5). p72^{syk} was also shown to be associated with the B cell antigen receptor (6) and another member of this PTK family, ZAP-70, is associated with the ζ chain of the TCR/CD3 complex in an activation-dependent manner (7).

As shown in several receptor-PTKs corresponding to such peptide growth factors as epidermal growth factor and platelet-derived growth factor, sequence motifs such as Src homology (SH) 2 and SH3, play pivotal roles in signal-transducing processes. SH2 and SH3 domains interact with phosphotyrosine in the context of surrounding sequences (8) and short proline-rich stretches (9), respectively. A bacterial fusion protein MBP-Lyn, composed of the maltose binding protein and Lyn, bound several tyrosine-phosphorylated proteins in activated rat basophils via the SH2 domain of Lyn. One of the Lyn binding proteins was a cytoskeletal protein, paxillin, suggesting a role of this interaction in the cytoskeletal changes that are prominent after mast cell activation (10).

In this chapter, we describe the structure and function of the recently discovered Tec family PTKs. Evidence for the involvement of the best characterized member of this family, Btk, in the FcεRI signaling will be presented. In this context the interaction between PTK and protein kinase C (PKC) via a sequence motif referred to as the pleckstrin homology (PH) domain seems to play a pivotal role. We also discuss the substrates for protein-tyrosine kinases (SPY) which may transduce the activation signal downstream of the activated PTKs.

1. Tec FAMILY PTKs

Although a large body of evidence has demonstrated essential roles of protein kinases in signal transduction, very few studies addressed how many and what kinds of protein kinases are expressed in a single cell. Since an optimal condition for one kinase differs significantly from that of another kinase and no segments of all known kinases are homologous enough to be recognized by a set of antibodies, the above question is conceptually quite difficult to answer. However, there is information which suggests that there is a relatively small number of PTKs expressed in a single cell. Some of protein kinases can be renatured on PVDF membranes after denaturation and detected by incubating with [γ-^{32}P]ATP and divalent cations. When this method was applied to mouse mast cells, we found twenty five protein kinase bands which were detected after KOH treatment and, therfore, probably represent PTKs (11). In a polymerase chain reaction (PCR) study on the mRNAs from K-562 chronic myelogenous leukemia cells, the presence of fourteen distinct PTK mRNAs was revealed (12).

Encouraged by the above observations we set out to find new PTK genes in mast cells. cDNAs synthesized from poly(A)+ RNA obtained from bone marrow-derived mouse mast cells (BMMC) were used as templates in two consecutive anchored PCRs. The PCRs were designed to amplify the putative 3' fragments of PTK cDNAs between the 3' ends of mRNAs and two degenerate oligonucleotides, corresponding to conserved PTK catalytic domain sequences, used as 5' nested primers. The PCR products were cloned and sequenced. Among seventy clones sequenced, two novel sequences, *emb* and *emt*, as well as six known kinase sequences, *lyn*, *hck*, *jak1*, *jak2*, IGF-I receptor and B-*raf*, were found (13). The open reading frame of the entire *emb* cDNA can encode the 660 residue sequence (Mr. 76,572). *emt* codes for the 619 amino acid sequence (Mr. 71,448). *emb* turned out to be the same gene referred to as *atk* (14) or *BPK* (15), which were isolated independently as the gene responsible for the defects in X-linked agammaglobulinemia patients. Later the same gene, now called by a unified term *btk* (Bruton tyrosine kinase), was demonstrated to be also mutated in an X-linked immunodeficient (*xid*) mouse (16,17). *emt* cDNA was isolated by others with the different names, *itk* (18), *Tsk* (19), or *emt* (20).

a. Structure of Btk and Emt

Overall primary structures of the Btk and Emt kinases show several notable features. First, they lack hydrophobic amino acid stretches characteristic of the transmembrane domains found in growth factor receptor PTKs. Second, they are similar to the Src family PTKs in displaying SH3, SH2 and a kinase domain that has a consensus autophosphorylation site corresponding to Tyr-416 in $p60^{c-src}$. Nevertheless, Btk and Emt differ from Src family PTKs in that they lack, first, a glycine residue at the second position that serves as a myristylation and membrane anchoring site and, second, a negative regulatory tyrosine residue (corresponding to Tyr-527 in $p60^{c-src}$) whose phosphorylation suppresses the kinase activity of Src family PTKs. These differences suggest a different mode of regulating the kinase activity of Emb and Emt relative to members of the Src family. Third, the kinase domains reveal relatively high homologies (55-64%) among Btk, Emt, TecII, a hematopoietic PTK (21), and Dsrc28 (22), a Src-related PTK isolated from *Drosophila melanogaster*. The homology level between any of these four kinases and other PTKs is significantly lower (<49%). Fourth, and more interestingly, Btk and Emt share a unique, extensively homologous amino-terminal region. Eighty eight of the 222 amino-terminal residues of Emb are also found within the first 177 residues of Emt. Basic amino acids are abundant in this region, *i.e.*, 41/222 residues in Emb and 33/177 residues in Emt. Within this region, the amino-terminal two-thirds form a PH domain (see below). Fifth, the Btk-specific sequence, PERQIPRRGEESSEME, which resides in the PH domain, has a high degree of homology (11 identities and 3 conserved substitutions over the 16-residue stretch) with a portion of the cytoplasmic domain of CD22 (23,24), also known as B lymphocyte cell adhesion molecule. This suggests that the same or similar protein(s) might interact with CD22 and Btk through this hydrophilic sequence motif. Finally, Btk and Emt share a unique proline-rich motif [(T/S)KKPLPPTPE(E/D)], found between the PH and SH3 domains, that has some homology with sequences in microtubule-associated proteins 2 and 4. Btk has another proline-rich motif [KKPLPPEP] seven residues downstream. These proline-rich sequences might interact with SH3 domains of other proteins. Based on the above considerations, Btk, Emt, TecII, and probably Dsrc28, appear to constitute a novel subfamily of PTKs that may perform similar functions, and share unique modes of regulation.

b. Phosphorylation and activation of Btk upon FcεRI cross-linking

Previous studies have shown that several proteins in the size range of 70-80 kilodaltons are phosphorylated on tyrosine residues (25,26). Since Btk is a 77-kDa PTK, we tested the possibility that Btk itself becomes phosphorylated on tyrosine. BMMC were passively sensitized with anti-dinitrophenyl (DNP) IgE and then stimulated by a multivalent antigen, DNP conjugates of human serum albumin (DNP-HSA), for various lengths of time. Precleared detergent lysates were immunoprecipitated with anti-Btk antiserum and immune complexes were subjected to immunoblot analysis with anti-phosphotyrosine mAb (4G10). The result demonstrated that the level of tyrosine phosphorylation of a 77-kDa protein increased markedly (>20 fold) from a low basal level after FcεRI cross-linking (27). The increase in tyrosine phosphorylation of this protein was detected in less than a minute and reached the maximal level around 1-3 min after stimulation. The identity of the 77-kDa protein as $p77^{btk}$ was confirmed by reprobing the same stripped blot with anti-Btk antibody.

We next metabolically labeled cellular proteins with ^{32}P and determined whether Btk is phosphorylated on other residues. The phosphorylation level of Btk in IgE/antigen-stimulated BMMC was higher than that in unstimulated cells. Phosphoamino acid analysis revealed an increase not only in tyrosine phosphorylation but also in serine and threonine phosphorylation (27).

As shown with many PTKs involved in signal transduction pathways for FcεRI as well as for T and B cell antigen receptors, tyrosine phosphorylation of PTKs is accompanied by their activation. We examined, therefore, the kinase activity of Btk in an in vitro immune complex kinase assay. Lysates prepared from unstimulated or antigen-stimulated BMMC were immunoprecipitated with anti-Btk antibody and immune complexes were incubated with [γ-^{32}P]ATP in the presence of an exogenous substrate, enolase. Reaction products were analyzed by SDS-PAGE and blotted. Autoradiography of the blot shows a significant increase (1.5-3 folds at its peak) in Btk-phosphorylating activity upon FcεRI cross-linking (27). The increase of Btk-phosphorylating activity was as rapid as its tyrosine phosphorylation, but lasted longer than the latter. Although the increase in the Btk-phosphorylating and enolase-phosphorylating activity was relatively small, it was consistently observed in five different Btk preparations. Since phosphoamino acid analysis indicated the presence of a co-precipitating serine/threonine kinase activity in Btk immune complexes, the extent of tyrosine phosphorylation of enolase by Btk was examined by probing the same blot with anti-phosphotyrosine mAb. The result shows that enolase tyrosine-phosphorylating activity in Btk immune complexes was remarkably increased by FcεRI cross-linking. The maximal activity (5-10 folds over the basal level) was attained around 3 min after antigen stimulation.

c. Btk localizations and FcεRI-stimulated translocation

The rapid activation of Btk suggests that it plays a role in membrane proximal events upon FcεRI cross-linking. In order to obtain further clues on the function of Btk in mast cells, its subcellular localization was determined. Unstimulated or antigen-stimulated BMMC were fractionated into nuclear, particulate (= membrane), and soluble (= cytosol) compartments, and Btk proteins in each fraction were detected by immunoblotting. Btk was found mostly (>90 %) in the cytosol in unstimulated cells while a small portion (≤5 %) was also found in the membrane fraction (27). The level of Btk in membrane fractions increased after antigen stimulation. The increase in membrane-associated Btk, detected in less than one minute, persisted for at least 15 min following receptor cross-linking. When the same blot was reprobed with anti-PKC(MC5) which reacts with α, β and γ isoforms, we observed that the amounts of membrane-associated PKC increased following FcεRI cross-linking. Interestingly, the kinetics of FcεRI-mediated increases in membrane-associated Btk and PKC were similar.

d. Btk is not associated with FcεRI

Both Src family PTKs, Lyn and c-Yes, and Syk PTK were shown to associate with FcεRI, suggesting their roles in the receptor-proximal events. We investigated the possibility of the Btk association with FcεRI by co-immunoprecipitation. To this end, unstimulated or IgE/antigen-stimulated BMMC or PT-18 cells were solubilized with mild detergents (10 mM CHAPS, 1% digitonin, or 0.5-1% NP-40) and immunoprecipitated with anti-Btk. Btk immunoprecipitates were analyzed by immunoblotting with anti-FcεRI β mAb or anti-FcεRI γ antibodies. We failed to detect any specific signals for these proteins. In reverse immunoprecipitations, anti-FcεRI β or anti-FcεRI γ immunoprecipitates were analyzed by immunoblotting with anti-Btk. Again the results were negative. Under the same conditions, anti-Lyn immunoprecipitates contained the proteins immunoreactive with anti-FcεRI β and anti-FcεRI γ. Another method to detect associated PTKs was also employed. Immune complexes precipitated with anti-FcεRI γ from CHAPS lysates of unstimulated or IgE/antigen-stimulated BMMC were subjected to in vitro kinase assay with [γ-^{32}P]ATP. ^{32}P-labeled proteins were released from antibodies, and re-immunoprecipitated with anti-Btk or anti-Lyn. The result showed that there were no Btk proteins in the anti-FcεRI γ immunoprecipitates while strong signals specific for Lyn (p56^{lyn} and p53^{lyn}) were clearly detected. Essentially identical results were obtained with in vitro kinase assays on anti-FcεRI β immunoprecipitates. Therefore, we conclude that Btk is not physically associated with FcεRI.

In a collaborative study we recently showed that Emt in T cells is associated with CD28 and tyrosine phosphorylated upon CD28 stimulation (28). CD28 is a major co-stimulatory receptor for T cell activation. Signaling via CD28 results in the stimulation of T cells to produce numerous lymphokines including interleukin-2 as well as the prevention of anergy induction. Therefore, it is reasonable to assume other Tec family kinases might be associated with a similar cell surface receptor of the immunoglobulin superfamily.

2. PH DOMAINS: BRIDGES BETWEEN PTK AND NON-PTK SIGNALING PATHWAYS ?

a. The PH domain of Btk binds directly to PKC in vitro.

The PH domain was originally recognized as repeat sequences in a prominent PKC substrate, pleckstrin (29,30). The homology search demonstrated the presence of the loosely conserved sequences of ~100 amino acids in numerous proteins. Some of them were found in signaling molecules such as protein serine/threonine or tyrosine kinases, GTPases, GTPase-activating proteins, guanine nucleotide exchange factors, and PLC-γ1 (31). Because PH domains were found in a PKC substrate, we tested the possibility that the PH domain of Btk may interact with PKC. Detergent lysates of BMMC or MCP-5 cells, an immortalized BMMC line, were incubated with immobilized GST fusion proteins containing the wild-type (GST-BtkPH) or *xid* (GST-BtkPH[*xid*]) PH domain of Btk. Analysis of proteins bound to the beads by immunoblotting with the monoclonal anti-PKC (MC5) showed that PKC bound to GST-BtkPH fusion proteins, but not to GST (32). Interestingly, three- to five-fold more GST-BtkPH(*xid*) proteins over the wild type counterpart were necessary to give comparable signals of bound PKC, suggesting that residue Arg-28 in the PH domain, which is substituted with cysteine in *xid* Btk, is involved in PKC binding. The GST fusion protein (GST-EmtPH) of the PH domain of Emt also bound to PKC in the same assay.

We determined whether the binding of PKC to the PH domain of Btk is direct or indirect. A mixture of highly purified rat brain PKC α, β and γ isoforms (>95% pure) were labeled with ^{32}P by autophosphorylation, and incubated with the blot retaining the purified GST or GST-BtkPH protein. GST-BtkPH proteins, but not GST, bound PKC (32), indicating that the PH domain of Btk directly bound PKC.

We next determined the Btk-interacting isoform in BMMC and MCP-5 by probing the blots retaining the proteins bound to GST-BtkPH beads with PKC isoform-specific antibodies. Both Ca^{2+}-dependent (α, βI, and βII) and Ca^{2+}-independent PKC isoforms (ε and ζ) bound to GST-BtkPH beads, while neither η nor θ, which are expressed in substantial amounts in BMMC and MCP-5, were detected among the proteins bound to GST-BtkPH beads. No detectable amount of PKC γ, or δ isoform was found in the total cell lysates of BMMC or MCP-5.

b. Btk is physically associated with PKC βI in mast cells.

In order to determine whether PKC is associated with Btk in mast cells, Btk in the lysates of unstimulated or IgE/antigen-stimulated BMMC was immunoprecipitated with anti-BtkC antibodies, which recognize the carboxyl-terminal twelve residues of Btk, and immune complexes were analyzed by immunoblotting with anti-PKC (MC5). The anti-PKC-reactive ~80-kDa band was clearly detected in both unstimulated and activated cells (32). Cross-linking of FcεRI did not affect the level of the co-precipitated PKC. Immunoblotting of the immune complexes with various isoform-specific antibodies revealed constitutive association of only PKC βI with Btk. In reciprocal experiments, Btk was detected in the anti-PKC (MC5) immunoprecipitates (32).

c. PKC phosphorylates and down-regulates Btk in vitro.

We tested the possibility that PKC phosphorylates Btk. To this end, the partially purified Btk from RBL-2H3 cells (33) or purified GST-BtkPH proteins were incubated with the purified rat brain PKC(α, β and γ) in the presence of PKC activators and [γ-^{32}P]ATP. PKC phosphorylated both Btk and GST-BtkPH proteins on serine residues (32).

Next, we examined the effects of phosphorylation by PKC on the ability of Btk for autophosphorylation. Btk partially purified from RBL-2H3 cells was incubated with cold ATP for 30 min at 30°C in the presence or absence of the purified rat brain PKC under PKC phosphorylation conditions. Btk in the mixtures was recovered by precipitation with anti-BtkC, and immune complexes were incubated with cold ATP. Tyrosine autophosphorylation of Btk was then detected by immunoblotting with anti-phosphotyrosine mAb. Phosphorylation of Btk by PKC resulted in a decrease in Btk autophosphorylation by about 60-80 % (32).

In order to confirm that the Btk-associated PKC can regulate the enzymatic activity of Btk by phosphorylation, PKC was immunoprecipitated with anti-PKC(MC5) from a mixture of purified rat PKC and partially purified Btk. Immune complexes were incubated at 30°C for 20 min with PKC activators in the presence or absence of cold ATP. After washing, the immune complexes were incubated at 25°C for 3 min with [γ-^{32}P]ATP under Btk autophosphorylation conditions, and reaction products resolved by SDS-PAGE and blotting were visualized by autoradiography. The result showed that Btk phosphorylation by the associated PKC inhibited the Btk autophosphorylating activity by 80-90% (32).

d. Effects of PKC modulation on tyrosine phosphorylation of Btk in response to FcϵRI cross-linking.

We investigated in vivo effects of PKC modulation on the FcϵRI-mediated Btk tyrosine phosphorylation. Chronic (18 hr) treatment of MCP-5 cells with 100 nM phorbol 12-myristate 13-acetate (PMA) decreased the level of anti-PKC(MC5)-reactive proteins to less than one-twentieth of that in untreated cells. Under this condition, tyrosine phosphorylation on Btk induced by FcϵRI cross-linking was higher and more rapid in PMA-treated cells than that in control cells (32). Augmentation of tyrosine phosphorylation of Btk subsequent to the IgE/antigen stimulation was also observed when MCP-5 cells were pretreated with PKC-specific inhibitors, calphostin C for 30 min or Ro31-8425 for 10 min (32).

Our interpretation of the negative regulation of Btk by PKC-mediated phosphorylation in the context of the FcϵRI signaling system is that Btk phosphorylation by PKC keeps in check the basal and peak tyrosine-phosphorylation levels of Btk. Indeed, the basal level of Btk tyrosine phosphorylation prior to antigen stimulation is higher in calphostin C (or RO31-8425)-treated cells than that in untreated cells. In support of this hypothesis, the Btk-PKC interaction is constitutive. Btk is heavily phosphorylated on serine residues before FcϵRI cross-linking and receptor cross-linking enhances not only tyrosine phosphorylation of Btk but also serine and threonine phosphorylation, although this data does not necessarily mean that the serine/threonine phosphorylation on Btk is due solely to its associated PKC.

Our preliminary data show that Emt is also phosphorylated on tyrosine, serine, and threonine residues and enzymatically activated upon FcεRI cross-linking. Association and enzymatic interaction between Emt and PKC isoforms in mast cells are also found. Therefore, the interaction between PKC and Tec family PTKs via their PH domains seems to be a universal phenomenon. This series of experiments established the biochemical basis for interactions between PTKs and PKC.

A recent report demonstrated that the carboxyl-terminal portion of various PH domains and their downstream sequences interact with the G-protein βγ subunits in vitro (34). This and our data, taken together, suggest that three important signaling systems, i.e., PTK, PKC and G-protein, may interact on the PH domain of Tec family PTKs. Further characterization on these interactions is under way.

3. SPY75, A SPY INVOLVED IN THE FcεRI SIGNALING SCHEME.

In order to better understand the FcεRI signaling pathway, we need to know more about the substrates phosphorylated by the PTKs discussed above. Several PTK substrates have been identified in mast cells and basophils which are activated via FcεRI (Table 1). They include the β and γ receptor subunits (35), phospholipase C (PLC)-γ1 (36), a protooncogene product, p95^{vav}(37), an SH2- and SH3-containing protein, Nck (38), and a MAP kinase, ERK2 (11). Phosphorylation of the β (on tyrosine and serine) and γ (on tyrosine and threonine) subunits of FcεRI takes place within 5 sec of receptor cross-linking and is restricted to cross-linked receptors. Thus, disengagement of the receptor accompanies almost instantaneous dephosphorylation of the β and γ subunits. This rapid and transient phosphorylation of activated receptors suggests that it may

Table 1. Known PTK substrates in FcεRI signaling pathway

PTK substrate	Mr (kDa)	Motifs
FcεRI β subunit	30	TAM
FcεRI γ subunit	14	TAM
Phospholipase C-γ	145	SH2 (2)[a], SH3, PH
Vav	95	cdc24[b], PH, DAG[c], SH3 (2)[a], SH2
SPY75 (=HS1)	75	37-residue repeat[d] (3.5)[a], SH3
Nck	47	SH2, SH3 (3)[a]
ERK2 (=p42^{mapk})	42	
Paxillin	65-70	

[a]Numbers in parentheses indicate frequencies of the motif in a molecule.
[b]guanine nucleotide exchange factor domain
[c]diacyl glycerol- and phorbol ester-binding sequence
[d]37-residue repeats with a putative helix-loop-helix structure

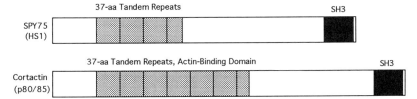

FIG. 2. Modular structures of SPY75 and cortactin.

serve to recruit specific components of the FcεRI signaling pathway. Tyrosine phosphorylation of PLC-γ1 by ligand (*e.g.*, epidermal growth factor)-activated growth factor receptor PTK appears to activate the enzymatic activity of PLC-γ1, accounting for an increased synthesis of inositol 1,4,5-trisphosphate (IP$_3$) and DAG. IP$_3$ mobilizes Ca^{2+} from intracellular storage sites and DAG activates PKC. In murine mast cells, PLC-γ1 is constitutively associated with a PTK (probably the 44-kDa protein), and FcεRI cross-linking appears to activate the latter enzyme (39). p95^{vav} has several sequence motifs, including one SH2, two SH3 regions, a putative diacylglycerol/phorbol ester binding domain, and a guanine nucleotide dissociation stimulator (GDS)-related sequence. Recently, the GDS function of p95^{vav} was shown to be regulated by tyrosine phosphorylation to activate the Ras protein in T and B cells (40). Another protein, Nck, composed almost exclusively of three SH3 domains and one SH2 domain, is tyrosine-phosphorylated upon FcεRI cross-linking. Phosphorylation of this potentially oncogenic protein is also induced by many other stimuli, including growth factors, cross-linking of the TCR, the B cell antigen receptor, and the low-affinity IgG receptor (FcγRII), and it is postulated that Nck functions as an adaptor molecule to link signals initiated by the receptor-coupled activated tyrosine kinases to downstream effectors in the signaling pathways for cell activation and growth. MAP kinases, also referred to as extracellular signal-regulated kinases (ERKs), are a family of serine/threonine kinases and are activated by chain phosphorylation reactions in the order of Raf (or MEKK)→MEK→MAP kinases in response to a variety of stimuli. MAP kinases, localized in both the cytoplasm and the nucleus, are thought to integrate the signals generated at the cell surface and transmit them to downstream effector sites by phosphorylating proteins such as c-Jun and S6 kinase to activate the transcription of certain genes and protein synthesis. In our recent study, a 75 kDa protein, termed SPY75, was identified as a major tyrosine-phosphorylated protein in activated mouse mast cells (41). SPY75, barely tyrosine phosphorylated in resting cells, was rapidly and transiently tyrosine phosphorylated upon FcεRI cross-linking in an antigen concentration-dependent manner. Similar SPY75 tyrosine phosphorylation was observed when antigen receptors on B and T lymphocytes were cross-linked by appropriate antibodies. However, IL-3, GM-CSF, or SCF did not induce tyrosine phosphorylation of SPY75 in PT-18 mast cells despite their responsiveness to these cytokines. This protein, the mouse homologue of the human *HS1* gene product, has putative repetitive helix-turn-helix motifs found in many DNA-binding proteins and a putative nuclear transport signal. It also has an SH3 domain which is found in many signaling molecules and cytoskeletal proteins. Similarities in the overall structures between SPY75 and cortactin (p80/85), a substrate for pp60^{v-src}, are salient (Fig. 2). Cortactin is localized in cortical structures such as membrane ruffles and lamellipodia in fibroblasts and binds to F-actin through its tandem repetitive sequences composed of 37 residues

(42). Therefore, SPY75 may have an actin-binding capacity and this potential propety, together with the carboxyl-terminal SH3 domain, may determine its subcellular localization. Another clue on the function of SPY75 was recently obtained in a collaborative study (43), which showed that SPY75 became associated with Ig-β, the *B29* gene product of the B cell antigen receptor complex when B104 human B cells were stimulated with anti-IgM or IgD. This result suggests another possibility that SPY75 may play a role in a membrane-proximal event in the FcεRI signaling pathway. These diferent, but not mutually exclusive, possibilities about the function of SPY75 must be tested. Despite the rapid progress in this field, there still remain many tyrosine-phosphorylated proteins to be identified. However, a recent study demonstrated that some of the tyrosine-phosphorylated proteins associated with the activated FcεRI have distinctive molecular masses (*i.e.*, multiubiquitinated β subunits of 42 kDa and 48 kDa, and multiubiquitinated γ subunits of at least six species of >14 kDa) due to multiubiquitination of the β and γ subunits of FcεRI (44). There are a lot to be learned about the interplay between the above-mentioned PTKs, PTK substrates, and other signaling molecules.

ACKNOWLEDGMENTS

Authors thank Drs. Teruko Ishizaka and Kimishige Ishizaka for their support and encouragement. Supported in part by NIH grant AI33617. This is Publication No. 109 from La Jolla Institute for Allergy and Immunology.

REFERENCES

1. Benhamou M, Gutkind JS, Robbins KC, Siraganian RP. Tyrosine phosphorylation coupled to IgE receptor-mediated signal transduction and histamine release. *Proc. Natl. Acad. Sci. USA* 1990;87:5327-5330.
2. Kawakami T, Inagaki N, Takei M, et al. Tyrosine phosphorylation is required for mast cell activation by FcεRI cross-linking. *J. Immunol.* 1992;148:3513-3519.
3. Eiseman E, Bolen JB. Engagement of the high-affinity IgE receptor activates *src* protein-related tyrosine kinases. *Nature* 1992;355:78-80.
4. Hutchcroft JE, Geahlen RL, Deanin GG, Oliver JM. FcεRI-mediated tyrosine phosphorylation and activation of the 72-kDa protein-tyrosine kinase, PTK72, in RBL-2H3 rat tumor mast cells. *Proc. Natl. Acad. Sci. USA* 1992;89:9107-9111.
5. Benhamou M, Ryba NJP, Kihara H, Nishikata H, Siraganian RP. Protein-tyrosine kinase p72^{syk} in high affinity IgE receptor signaling. Identification as a component of pp72 and association with the receptor γ chain after receptor aggregation. *J. Biol. Chem.* 1993;268:23318-23324.
6. Hutchcroft JE, Harrison ML, Geahlen RL. Association of the 72-kDa protein-tyrosine kinase PTK72 with the B cell antigen receptor. *J. Biol. Chem.* 1992;267:8613-8619.
7. Chan AC, Iwashima M, Turck CW, Weiss A. ZAP-70: a 70-kd protein-tyrosine kinase that associates with the TCR ζ chain. *Cell* 1992;71:649-662.

8. Koch CA, Anderson D, Moran MF, Ellis C, Pawson T. SH2 and SH3 domains: Elements that control interactions of cytoplasmic signaling proteins. *Science* 1991;252:668-674.
9. Ren R, Mayer BJ, Cicchetti P, Baltimore D. Identification of a ten-amino acid proline-rich SH3 binding site. *Science* 1993;259:1157-1160.
10. Minoguchi K, Kihara H, Nishikata H, Hamawy MM, Siraganian RP. Src family tyrosine kinase Lyn binds several proteins including paxillin in rat basophilic leukemia cells. *Mol. Immunol.* 1994; in press.
11. Fukamachi H, Takei M, Kawakami T. Activation of multiple protein kinases including a MAP kinase upon FcεRI cross-linking. *Int. Arch. Allergy Immunol.* 1993;102:15-25.
12. Partanen J, Makela TP, Alitalo, R, Lehvaslaiho H, Alitalo K. Putative tyrosine kinases expressed in K-562 human leukemia cells. *Proc. Natl. Acad. Sci. USA* 1990;87:8913-8917.
13. Yamada N, Kawakami Y, Kimura H. et al. Structure and expression of novel protein-tyrosine kinases, Emb and Emt, in hematopoietic cells. *Biochem. Biophys. Res. Commun.* 1993;192:231-240.
14. Vetrie DJ, Vorechovsky I, Sideras P, et al. The gene involved in X-linked agammaglobulinemia is a member of the *src* family of protein-tyrosine kinase. *Nature* 1993;361:226-233.
15. Tsukada S, Saffran DC, Rawlings DJ, et al. Deficient expression of a B cell cytoplasmic tyrosine kinase in human X-linked agammaglobulinemia. *Cell* 1993;72:279-290.
16. Thomas JD, Sideras P, Smith CIE, Vorechovsky I, Chapman V, Paul WE. Colocalization of X-linked agammaglobulinemia and X-linked immunodeficiency genes. *Science* 1993;261:355-358.
17. Rawlings DJ, Saffran DC, Tsukada S, et al. Mutation of unique region of Bruton's tyrosine kinase in immunodieficient XID mice. *Science* 1993;261:358-361.
18. Siliciano JD, Morrow TA, Desiderio SV. *itk*, a T-cell-specific tyrosine kinase gene inducible by interleukin 2. *Proc. Natl. Acad. Sci. USA* 1992;89:11194-11198.
19. Heyeck SD, Berg LJ. Developmental regulation of a murine T-cell-specific tyrosine kinase gene, *Tsk*. *Proc. Natl. Acad. Sci. USA* 1993;90:669-673.
20. Gibson S, Leung B, Squire JA, et al. Identification, cloning, and characterization of a novel human T-cell-specific tyrosine kinase located at the hematopoietin complex on chromosome 5q. *Blood* 1993;83:1561-1572.
21. Mano H, Mano K, Tang B, et al. Expression of a novel form of *Tec* kinase in hemapoitic cells and mapping of the gene to chromosome 5 near *kit*. *Oncogene* 1993;8:417-424.
22. Gregory RJ, Kammermeyer KL, Vincent WS III, Wadsworth SG. Primary sequence and developmental expression of a novel *Drosophila melanogaster src* gene. *Mol. Cell. Biol.* 1987;7:2119-2127.
23. Stamenkovic I, Seed B. The B-cell antigen CD22 mediates monocyte and erythrocyte adhesion. *Nature* 1990;345:74-77.
24. Wilson GL, Fox CH, Fauci AS, Kehrl JH. cDNA cloning of the B cell membrane protein CD22: a mediator of B-B cell interactions. *J. Exp. Med.* 1991;173:137-146.
25. Benhamou M, Stephan V, Robbins KC, Siraganian RP. High-affinity IgE receptor-mediated stimulation of rat basophilic leukemia (RBL-2H3) cells

induces early and late protein-tyrosine phosphorylations. *J. Biol. Chem.* 1992;267:7310-7314.
26. Minoguchi K, Benhamou M, Swaim WD, Kawakami Y, Kawakami T, Siraganian RP. p72^{syk} is the major protein tyrosine kinase but a minor component of pp72, the tyrosine phosphorylated proteins prominent in FcεRI-activated cells. *J. Biol. Chem.* 1994; in press.
27. Kawakami Y, Yao L, Tsukada S, Witte ON, Kawakami T. Tyrosine phosphorylation and activation of Bruton tyrosine kinase upon FcεRI cross-linking. *Mol. Cell. Biol.* 1994; in press.
28. August A, Gibson S, Mills GB, Kawakami Y, Kawakami T, Dupont B. CD28 is associated with and induces the immediate tyrosine phosphorylation of the *tec* family kinase *itk/tsk/emt*. *Proc. Natl. Acad. Sci. USA* 1994; in press.
29. Haslam RJ, Koide HB, Hemmings BA. Pleckstrin domain homology. *Nature* 1993;363:309-310.
30. Mayer BJ, Ren R, Clark KL, Baltimore D. A putative modular domain present in diverse signaling proteins. *Cell* 1993;73:629-630.
31. Musacchio A, Gibson T, Rice P, Thompson J, Saraste M. The PH domain: a commom piece in the structural patchwork of signalling proteins. *Trends Biochem. Sci.* 1993;18:343-348.
32. Yao L, Kawakami Y, Kawakami T. The pleckstrin homology domain of Btk tyrosine kinase interacts with protein kinase C. *Proc. Natl. Acad. Sci. USA* 1994; in press.
34. Touhara K, Inglese J, Pitcher JA, Shaw G, Lefkowitz RJ. Binding of G protein βγ-subunits to pleckstrin homology domains. *J. Biol. Chem.* 1994;269:10217-10220.
33. Price DJ, Kawakami Y, Kawakmi T, Rivnay B. Purification of a major tyrosine kinase from RBL 2H3 cells phosphorylating FcεRI γ-cytoplasmic domain and identification as the *btk* gene product. submitted for publication.
35. Paolini R, Jouvin M-H, Kinet J-P. Phosphorylation and dephosphorylation of the high-affinity receptor for immunoglobulin E immediately after receptor engagement and disengagement. *Nature* 1991;353:855-858.
36. Park DJ, Min HK, Rhee SG. IgE-induced tyrosine phosphorylation of phospholipase C-γ1 in rat basophilic leukemia cells. *J. Biol. Chem.* 1991;266:24237-24240.
37. Margolis B, Hu P, Katzav S, et al. Tyrosine phosphorylation of *vav* proto-oncogene product containing SH2 domain and transcription factor motifs. *Nature* 1992;356:71-74.
38. Li W, Hu P, Skolnik EY, Ullrich A, Schlessinger J. The SH2 and SH3 domain-containing Nck protein is oncogenic and a common target for phosphorylation by different surface receptors. *Mol. Cell. Biol.* 1992;12:5824-5833.
39. Fukamachi H, Kawakami Y, Takei M, Ishizaka T, Ishizaka K, Kawakami T. Association of protein-tyrosine kinase with phospholipase C-γ1 in bone marrow-derived mouse mast cells. *Proc. Natl. Acad. Sci. USA* 1992;89:9524-9528.
40. Gulbins E, Coggeshall KM, Baier G, Katzav S, Burn P, Altman A. Tyrosine kinase-stimulated guanine nucleotide exchange activity of Vav in T cell activation. *Science* 1993;260:822-825.

41. Fukamachi H, Yamada N, Miura T, et al. Identification of a protein, SPY75, with repetitive helix-turn-helix motifs and an SH3 domain as a major substrate for protein tyrosine kinase(s) activated by FcεRI cross-linking. *J. Immunol.* 1994;152:642-652.
42. Wu H, Parsons JT. Cortactin, an 80/85-kilodalton pp60^{src} substrate, is a filamentous actin-binding protein enriched in the cell cortex. J. Cell Biol. 1993;120:1417-1426.
43. Hata D, Nakamura T, Kawakami T, Kawakami Y, Herren B, Mayumi M. Tyrosine phosphorylation of MB-1, B29, and HS1 mproteins of the human B cell following receptor crosslinking. *Immunol. Lett.* 1994; in press.
44. Paolini R, Kinet J-P. Cell surface control of the multiubiquitination and deubiquitination of the high-affinity immunoglobulin E receptors. *EMBO J.* 1993;12:779-786.

Subject Index

A

Acetylcholine, 175
Acetylsalicylic acid, 197
Adenosine triphosphate
  in 5-LO synthesis, 29
  IL-3 effects in, 54
β-Adrenergic agonists, 196
Allergic diseases/reactions
  anti-IgE autoantibodies in, 186
  basophil action in, 51–52, 161, 183, 193–194, 202
  bee-sting anaphylaxis, 167, 168
  in helminth immunity, 212
  inflammatory processes, 51–52
  integrin-mediated signals in, 240
  mast cell response in, 131, 161, 183, 193–194
  stem cell factor in, 8
  tryptase activity in, 163
Anaphylaxis, 167, 168, 216–218
Antigen receptor activation motif, 249
AP1, 44, 45, 47
Apoptosis suppression
  cytokines in, 52
  stem cell factor in, 2, 3–4
Arachidonic acid
  IL-3 mediation, 55–56
  LC$_4$ synthase and, 25, 30
  5-LO and, 25, 29–30
  in mast cells, 27–30
  metabolism, 25
  phospholipase A$_2$ and, 25, 27–28
  prostaglandin D$_2$ and, 29
  prostaglandin endoperoxide synthase and, 25, 28–29
Asthma, 193
  corticosteroids in, 197–198
  non-steroidal anti-inflammatory drugs in, 196–197
  tryptase activity in, 168
Autocrine cytokine loops, 109, 113, 114

B

Basophils
  in allergic/inflammatory processes, 51–52, 161, 183, 193–194, 202
  cellular mediators, 121–122
  characteristics, 51
  corticosteroid response, 198
  cytokine interactions, 120, 123–125
  development, 119–120, 123–125
  differentiation antigens, 122
  distribution, 194
  function, 119
  in helminth infection, 211
  histamine release in urticaria, 184–186
  identification in asthma, 193
  IL-4 mRNA production, 58
  IL-3 regulation of, 51–56
  in IL-4 release, 52, 56–60
  immunophilins and, 201–202
  malignant cells, 125–126
  negative regulators, 123
  nerve growth factor regulation, 121
  NSAID response, 197
  phosphodiesterase inhibitors and, 196
  protease activity, 161
  protein kinase C activity in, 198
  tyrosine kinases in, 199
  vs. mast cells, 92–95, 194
Bcl-2, 4
Bee-sting anaphylaxis, 167, 168
BPO$_2$/BPO-lysine, 59
Bradykinin, tryptase and, 163
Bryostatins, 198–199
Bt2cAMP, 41–42
Budesonide, 198

C

c-*fos*, 81–82, 83, 84, 85
c-*kit* protein
  activation mutation, 99–100
  in hematopoiesis, 87
  in mast cell development, 93, 95–98
  in mast cell survival, 98–99
  tyrosine kinase domain, 68, 92, 95–98
c-*myc* gene activation, 81, 82–83

263

Calcineurin, 200, 201
Calcium-dependent membrane-binding domain, 27
Calcium metabolism
 in arachidonic acid metabolism, 27
 in histamine desensitization events, 59–60
 in histamine release, 226–229, 234
 in IL-3 mediation of basophil, 54, 55–56
 in interleukin-4 release, 60
 5-LO and, 29–30
 in mast cell adhesion to laminin, 140
 in Th cytokine expression, 40–41
Cardiopulmonary response, chronic SCF treatment and, 6
Catecholamines, 196
CD4, 39, 40
CD8, 39, 40
CD28, 39, 253
CD34+ cells, 19–21, 119, 140
Chronic myeloid leukemia, 125
Chymase, 123, 166
 mast cell granule, in helminthosis, 212–213, 215, 222
Ciliary neurotrophic factor, 67, 80–81
Collagen, 240–241, 245
Compound 48/80, 226, 229, 234
Concanavalin A, 226
Conserved lymphokine element 0 (CLEO), 44–45
Corticosteroids, 197–198
CSF-1, 67
Cyclic adenosine monophosphate (cAMP)
 in FMLP release, 53
 in inflammation, 194–195
 interleukin-2 effects, 229, 234
 in interleukin transcription, 41–42
 phosphodiesterase isozymes, 195–196
Cyclins/cyclin dependent kinases, 83
Cyclooxygenase. *See* Prostaglandin endoperoxide synthase
Cyclophilin, 199–201
Cyclosporin
 applications, 199
 cytokine inhibition by, 41
 immunosuppressive action, 200–202
 in mast cell protease secretion, 215
 in signal transduction, 199
 in urticaria treatment, 190
Cytokines. *See also* Interleukin factors; *specific factor*
 in allergic inflammation, 51–52
 autocrine loops, 109, 113, 114
 in basophil development, 92, 120, 123–125
 in basophil regulation, 52
 CLEO in differential regulation of, 44
 DNA motifs for regulation of, 41
 fibronectin receptor integrins in expression of, 243–244
 function, 79
 hematopoietic receptor network, 67–68, 87
 interactions in signal transduction, 66–67, 73–74, 87
 in keratinocytes, 133
 in mast cell adhesion, 141
 in mast cell development, 123–125, 225
 in mast cell protease secretion, 214–215
 in mast cell regulation, 65
 mast cell specificity, 67
 in neuronal growth, 177–178
 nuclear factor of activated T cells in regulation of, 44–47
 production, 79, 133
 in prostaglandin D_2 generation, 30–33
 receptor superfamily, 67–68, 80–81
 in regulation of $p21^{ras}$, 70–73, 75
 in regulation of Shc protein, 71
 response determinants, 79–80
 Sos1 and, 72–73
 Th cell expression, 39–41
 Th cell specificity in expression of, 42–44
 tyrosine kinase receptors on mast cells, 67

D
Dermatitis, 193

E
Eicosanoid biosynthesis, 25, 27–28
Embryonic stem cell cultures, 112–114
Eosinophils
 in allergic reactions, 51
 in helminth infection, 211
 phosphodiesterase inhibitor-modulated, 196
 production, 39
Erythropoietin, 67
Exon/intron organization, 151, 165

F
FcεRI
 Btk and, 252–253
 expression on mast cell-committed progenitors, 109–112, 114
 histamine release in urticaria and, 186–189, 190
 IL-3 expression, 111, 112, 113
 in mast cell adhesion, 140, 141, 143

in mast cell signal transduction, 5, 25, 239–240
NSAID response, 197
PKC modulation of Btk phosphorylation and, 255–256
p72^{syk} and, 250
PTK in mast cell activation through, 249–250
SPY75 and, 256–258
structure, 5, 249
Fenoterol, 196
Fibronectin, 240–241
  mast cell adhesion, 140–141, 143
  mast cell survival on, 245
  receptor integrins, 243–244, 246
  tryptase activity, 163
Fibrinogen, 240–241
FK-binding proteins, 199–201, 202
FLAP, 30
Fluticasone, 198
FMLP, 55, 60
  vs. immunoglobulin-E, 52–53
Formoterol, 196
Forskolin, 197

G

Gensitein, 141
Germ cells, c-*kit* mRNA and, 99
gp130, 19
Granule phenotype, 154–155
Granulocyte-macrophage colony stimulating factor, 44–45, 67, 225, 244
  in basophil development, 120
  c-*myc* gene activation, 81, 82–83
  chimeric receptor, 84–85
  GMR subunit functioning, 81–82, 83, 85
  in neuronal growth, 178
  in p21^{ras} induction, 70–71
  receptor superfamily, 80–81
Growth factor receptors, 87

H

Helminth infection
  acquired immunity, 211, 212
  immunoglobulin E and, 211
  mast cell hyperplasia in, 212, 219
  mast cell proteases in, 212–213, 215–218, 222
  mucosal permeability in, 215–218
  rapid expulsion phenomenon, 211–212, 219
  stem cell factor in, 218–219, 222
Hematopoietic system, 8, 67–68, 79, 87
Heparin, tryptase and, 162
Herbimycin, 54–55

Histamine
  autoantibody release in urticaria, 184–189
  basophil metabolism and, 51
  calcium metabolism in release of, 226–229, 234
  as cell mediator, 121–122
  corticosteroid inhibition of, 225
  cytokine inhibition of, 225, 226
  desensitization events, 59–60
  in electrical short circuit current response, 175
  IL-2 inhibition of, 232, 234–235
  IL-mediated release, 52, 53, 55–56, 58–59
  immunophilin-mediated, 201–202
  interleukin-2 inhibition of, 226, 232–235
  [^{3}H]-leucine uptake and, 229, 234
  mast cell adhesion in release of, 143
  NSAID response, 197
  phosphodiesterase inhibitor-modulated, 196
  protein kinase A inhibition in, 229, 234
  protein kinase C activity in, 198
  sera-induced release in urticaria, 184–186
  therapeutic implications in urticaria, 190
  tyrosine kinases and, 199, 232, 234–235
  in urticaria, 183–184
Hyperplasia, mast cell
  in helminthosis, 212
  SCF-induced, 3–4, 7–8

I

Immortalized mast cell lines, 154–155
Immune system
  basophil/mast cell pharmacology, 193–202
  chronic urticaria as disease of, 187, 189–190
  immunophilin ligands in, 199–202
  nervous system interaction, 173–174
Immunoglobulin-E
  in allergic reactions, 193–194
  basophil metabolism, 51, 53–54
  in helminth infections, 211
  in histamine release in urticaria, 184–189
  interleukin-4 and, 58–59
  in mast cell colony formation, 110
  protein kinase C and, 198–199
  in stem cell factor activity, 8
  vs. FMLP, 52–53
Immunoglobulin-G, 186
Immunophilins, 199–200
Indomethacin, 196–197
Interferons, 66
  α/β, 65
  in basophil development, 123
  in mast cell development, 66, 123

Interferons (contd.)
  in mast cell regulation, 65
Interleukin-1, 52
  β, 26, 32
  in histamine release, 225, 226
  nerve growth factor and, 177–178
Interleukin-2, 41, 42–44, 45, 81
  in ^{45}Ca uptake, 226–229, 234
  in histamine inhibition, 232–235
  in histamine release, 225, 226
  in IP$_3$ production, 226–229, 234
  [^{3}H]-leucine and, 229, 234
  in lipocortin formation, 231–232, 234–235
  protein synthesis elicited by, 230
  shared receptor system, 85–86
  tyrosine kinase activity, 232
Interleukin-3, 26, 32, 39, 41, 244
  in basophil generation of IL-4, 57, 60
  in basophil regulation, 51–56, 120
  c-*myc* gene activation, 81, 82–83
  clinical applications, 52
  Fc receptor expression, 11, 112, 113
  GMR receptor system, 80
  GMR subunit functioning, 81–82, 83
  in histamine release, 225, 226
  identification, 65, 154
  IL-4 interactions, 66
  incubation duration and effects of, 53–56
  in mast cell development, 13–14, 65–66
  mast cell proteases and, 154, 214–215
  in mast cell survival, 245
  in neuronal growth, 178
  phosphorylation effects, 54–55, 66
  in regulation of p21^{ras}, 70–71
  in Sos1 function, 72–73
  in stem cell factor effects, 2
  steroid effects, 53
  in tyrosine phosphorylation, 68, 74, 75
Interleukin-4, 14–15, 26, 39, 41, 65
  in basophil development, 121
  basophil release, 52, 53, 56–60
  [CA^{++}]$_i$ and, 60
  fibronectin receptor integrins and, 244
  histamine release and, 58–59, 225, 226
  IgE-mediated reactions and, 58–59
  IL-3 interactions, 66
  IL-2 receptor system and, 85–86
  in mast cell differentiation, 66
  mast cell proteases, 154
  mRNA production in basophils, 58
  in regulation of p21^{ras}, 70
  in tyrosine phosphorylation, 70, 73, 74, 75
Interleukin-5, 41, 42–44, 52, 81

  in basophil development, 120
  GMR α subunit, 83
  GMR receptor system, 80
  in histamine release, 226
Interleukin-6, 17–21, 32, 41, 176
  in neuronal growth, 178
  receptors, 80–81
Interleukin-7, 81
Interleukin-8, 52, 121
Interleukin-9, 26, 32, 121, 123, 154
  identification, 65
Interleukin-10, 16–17, 26, 65, 121, 123, 154
  in cytokine-induced PGD$_2$ generation, 32, 33
Interleukin-11, 17–20, 81
Interleukin-13, 73
Interleukin factors. *See also specific factor*
  in basophil regulation, 51–52
  Bt2cAMP effects on transcription, 41–42
  in histamine release, 225–226
  in KL-induced prostaglandin D$_2$ generation, 30–33
  in mast cell development, 26, 65–66, 121
  mast cell hemopoietin receptors, 67–68
  receptor superfamily, 80–81
  SCF combinations in mast cell growth, 16–21
  synergistic interactions, 66–67, 120
  Th cell expression, 39–41
IP$_3$, 226–229, 234
Isoproterenol, 196

K

Keratinocytes
  cytokine secretion, 133
  mast cell growth factor in, 137
Kit ligand growth factor. *See* Stem cell factor
KL
  in arachidonic acid metabolism in BMMC, 26
  in prostaglandin D$_2$ generation, 30–33

L

Laminin, 240–241
  mast cell adhesion, 139–140, 143
Lavendustin A, 199
[^{3}H]-Leucine, 229, 230, 234
Leukemia, 125–126
Leukemia inhibitory factor, 67, 81, 177, 178
Leukocytes. *See also* Basophils
  in allergic reactions, 51
  in emotional responses, 173
Leukotrine C$_4$
  histamine and, 56
  in IL-3 regulation of basophil function, 52, 53–55

immunophilin-mediated release, 201–202
NSAID response, 197
phosphodiesterase inhibitor-modulated, 196
production, 25, 31
structure, 30
Lipocortin formation, 225, 231–232, 234–235
5-Lipoxygenase (5-LO), 25, 29–30

M

Major histocompatibility complex, 39, 65, 66
MAP-kinases, 5, 27, 68, 70, 72, 256, 257
Mast cell adhesion
  to ECM proteins, 240–243
  to fibronectin, 140–141, 143, 240–241
  integrin molecules in, 241–243
  to laminin, 139–140, 143, 240–241
  processes, 141–143
  to vitronectin, 140–141, 143, 240–241
Mast cell-committed progenitors
  in bone marrow, 106–107, 109, 114
  definition, 105, 114
  laminin adhesion, 140
  in mast cell development, 105–106, 119–120
  peritoneal, 108–109
  retroviral transformation, 154–155
Mast cell degranulation
  anaphylactic type, 7
  chronic SCF treatment and, 5–7
  FcεRI in, 5
  histamine/tryptase release in, 165–166
  in neurostimulation, 175–177, 178
  SCF-induced, 4–5
Mast cell development
  in bone marrow, 119
  c-*kit* receptor in, 93, 95–98
  cellular mediators, 121–122
  committed progenitors in, 105–109, 119–120
  from cord blood, 17–21
  cytokine interactions in, 14–21, 66–67, 73–74
  cytokines in, 25–26, 65, 123–125, 225
  differentiation antigens, 122
  Fc receptors in, 109–112, 114
  hyperplasia in helminthosis, 212, 219
  in infection, SCF and, 219–221
  interleukin-3/4 in, 13–16, 65–66
  malignant, 125–126
  negative regulators, 123
  neoplastic transformation, c-*kit* in, 99–100
  nerve growth factor in, 177
  precursors, 13, 94–95
  protein-tyrosine kinases in, 249
  in squamous cell carcinoma, 132–137

stem cell factor in, 1, 2–3, 8–9, 14–16, 65–66, 93
stroma cell-derived growth factors in, 120–121
survival of, c-*kit* protein in, 98–99
in tumors, 131–132
vs. basophil development, 92
Mast cell distribution, 139, 194
  in nervous sytem, 174–175
Mast cell function, 119
  in disease, 131
  in electrical short circuit current response, 175
  in helminth infection, 211, 221–222
  matrix interactions, 143–145, 240
  nervous system functioning and, 175–178
  in pathogenesis, 183, 190, 193–194, 202
  phosphodiesterase inhibitor-modulated, 196
  signal transduction, 239
  in urticaria, 183
Mast cell growth factor. *See* Stem cell factor
Mast cell leukemia, 125–126
Mast cell proteases, 161. *See also* Tryptase
  amino acid sequence, 151
  baseline secretion, 214–215
  as cellular markers, 165–166
  exon/intron organization, 151
  function, 149
  gene clustering, 150–151
  genetic structure, 150–151
  granule packaging, 152–153
  granule phenotype determination, 154–155
  in helminthosis, 212–213, 215–218, 222
  heterogeneity, 153
  interleukin factors and, 152
  products, 149
Mast cell secretory function
  in allergic reaction, 51, 161
  corticosteroid response, 198
  cyclosporin and, 201
  cytokines in, 219
  in neuronal growth, 177
  NSAID response, 197
  protein kinase C in, 198–199
  stem cell factor in, 4–5, 8
  tyrosine kinases in, 199
Mast cells
  arachidonic acid metabolism in, 27–30
  basophils and, 51
  bone-marrow, 13–17, 25, 26, 66, 139
  Btk-PKC interactions, 254–256
  ^{45}Ca uptake, 226–229, 234

Mast cells (contd.)
  connective tissue-type, 13, 14, 25, 26, 66, 92, 93, 95–98, 225, 234
  cytokine-induced prostaglandin D_2 generation in, 30–33
  cytokine-induced tyrosine phosphorylation, 68–70
  cytokine receptors, 67–68
  embryonic stem cell-derived, 112–114
  fibronectin integrins in, 243–244, 246
  heterogeneity, 123
  [^{3}H]-leucine uptake, 229, 234
  lipocortin formation in, 225
  morphology, 13, 139
  mucosal, 13, 25, 66, 92, 93, 177, 211–212
  murine deficiency, 91
  p170 cytoplasmic protein in, 73
  peritoneal, 108–109, 226
  receptor systems, 87
  tryptase as clinical indicator, 166–168
  vs. basophils, 92–95, 194
Mastocytosis, 167, 168
Meclofenamic acid, 197
Mesenteric lymph node, neurite growth in, 177–178
MK-886, 30
mpl-Ligand, 67
mRNA
  c-kit, 98–99, 100
  interleukin-4, basophil expression, 58

N
Nck protein, 257
Nerve growth factor, 121, 123, 134
  IL-1 response, 177–178
  in mast cell growth, 177
Nervous system
  immune cells in, 176–178
  immune system interaction, 173–174
  mast cell degranulation in, 175–177, 178
  mast cell distribution in, 174–175
  structure, 174
Neuropeptides
  in immune response, 174
  tryptase degradation, 163
Neutrophil, 196
NGF, 33
Nimesulide, 197
Nippostrongylus brasiliensis, 92–95, 107, 108, 177, 215, 219–221
Non-steroidal anti-inflammatory drugs, 196–197
  action, 29

NS-398, 29
Nuclear factor of activated T cells, 44–47

O
Oncastatin M, 67, 80–81

P
p170 cytoplasmic protein, 73, 75
Pentoxifylline, 196
PH domains, 254–256
Pharmacology
  corticosteroids, 197–198
  cyclosporins, 199–202
  immunoregulatory cytokines, 202
  mediator release, 194
  non-steroidal anti-inflammatory drugs, 29, 196–197
  phosphodiesterase inhibitors, 194–195
  protein kinase C activity, 199
Phosphodiesterase inhibitors, 194–195
Phospholipase A_2 (PLA_2)
  in arachidonic acid metabolism, 25, 27–28, 55
  distribution, 27
  in inflammation, 27–28
  KL-stimulated expression, 31
  subclasses, 27, 55
Phospholipases, 234–235, 256
Platelet-derived growth factor, 67, 70
Pleckstrin homology, 250, 254
p21^{ras}
  cytokines in regulation of, 70–71, 75
  function, 70
  Shc and, 40, 71
  Sos1 activity and, 72–73
Priming
  agents, 175–176
  IL-3, 52, 53, 54, 57
  IL-4, 57–58
  KL, 26
  for PGD_2 generation, 30–31, 33
  steroids, 53
Prostaglandin D_2
  in connective tissue-type vs. mucosal mast cells, 25
  cytokine induced generation, 30–33
  IgE-dependent generation, 30–31
  NSAID response, 197
  synthase enzymes, 25, 29
Prostaglandin E2, 41
Prostaglandin endoperoxide synthase
  in arachidonic acid metabolism, 25, 28–29
  cytokine-stimulated expression, 31–33

SUBJECT INDEX

NSAIDs and, 29
structure, 28
Protein kinase A, 42
  interleukin-2 effects, 229, 234
Protein kinase C, 40–41, 54, 239
  Btk interaction, 254–256
  in p21^{ras} induction by cytokine, 70
  PTK interactions, 250
  in signal transduction, 198
Protein tyrosine kinases
  Btk and FcεRI, 252–253
  Btk PH domain, 254
  Btk structure, 251–252
  Emt, 251–252, 253, 256
  p72^{syk}, 250
  in signal transduction, 250
  Src family, 39, 250
  structure, 250
  substrate SPY75, 250, 256–258
  Tec family, 250–251
4PS cytoplasmic protein, 73
Psoriasis, 183

R

Rapamycin, 5, 200, 201, 202
Retroviral transformation, 154–155
Rolipram, 195, 196

S

Salmeterol, 196
Scleroderma, 163
Serglycin expression, 152–153
Shared receptor systems, 80–81, 85–86, 120
Shc protein, 40, 71, 75
Sos1, 72–73, 75
Splenocytes, 177
SPY75, 250, 256–258
Squamous cell carcinoma
  mast cell accumulation in, 131–132
  mast cell growth factors, 133–137
STAT transcription factors, 74
Staurosporine, 53, 54, 140, 198, 199
Steel factor. *See* Stem cell factor
Stem cell factor
  action, 121
  in apoptosis suppression, 2, 3–4
  chronic treatments, 5–7
  chymase formation, 123
  clinical application, 8–9
  development, 120–121
  in experimental primates, 7
  function, 1, 67
  in helminth infection, 212, 218–219, 222

in humans *in vivo*, 7–8
identification, 1–2, 14, 65
in immunoglobulin-E-dependent anaphylaxis, 5–7
interleukin factor combinations in mast cell growth, 16–21
in mast cell adhesion, 141–143
in mast cell colony formation, 110
in mast cell development and survival, 2–3, 8, 14–16, 26, 65–66, 93
in mast cell development in infection, 219–221
in mast cell secretory function, 4–5, 8
in regulation of p21^{ras}, 70–71
in Sos1 function, 72–73
tryptase formation, 123
tyrosine kinase receptors, 67
tyrosine phosphorylation induced by, 68, 75
Steroids, 53
Stroma cell-derived growth factors, 120–121
Substance P, 174, 175–176, 178
Survival, mast cell
  c-*kit* in, 98–99
  on fibronectin, 245
  stem cell factor in, 2–3, 8–9, 14–16, 65–66

T

T/TC cells, 21, 123, 253
TCRζ, 39–40
TCR phosphorylation, 39–41
TGF-β, 33, 120
Th cells
  CLEO and, 44–45
  cytokine expression in, 16, 39–41
  differentiation, 39
  nuclear factor of activated T cells, 44–47
  subset-specific expression, 42–44
Theophylline, 196
[^{3}H]Thymidine uptake, 135
Thymocytes, 177
TNF-α, 33, 244
TPA, 198–199
Tryptase
  biological activity, 163
  dissociation, 162
  exon/intron organization, 165
  in mast cell, 161
  as mast cell marker, 165–166
  molecular biology, 164–165
  pathological assessment, 166–168
  physical and enzymatic properties, 161–162
Tryptase/chimase mast cells, 13, 21, 123
Tumors. *See also* Squamous cell carcinoma

Tumors (contd.)
    mast cell accumulation, 131–132
    mast cell growth in, 132
    mast cell migration in, 135
Tyrosine activation motif, 249
Tyrosine kinase, 67
    c-*kit* point mutations, 95–98
    in histamine release, 199
    in IL-2 inhibition of histamine, 232, 234–235
    in mast cell activation, 249
    in signal processes, 199
Tyrosine phosphorylation
    Btk, 252, 255–256
    c-*kit*, 99
    cytokine-induced, 68–70, 75
    FcεRI signaling pathway and, 256–257
    IL-3 in, 54–55, 66, 74, 75
    IL-4 in, 73, 74
    mast cell adhesion in, 143
    in p21ras regulation, 70–71
    of Shc protein, 71
    in signal transduction, 239, 249–250

U

Urticaria
    clinical model, 183, 184, 193
    histamine releasing autoantibodies in, 184–189
    skin reaction to autologous serum, 184
    therapeutic implications of histamine activity, 187, 189–190
    tryptase and, 163

V

V3 cells, 155
Very late activation antigen, 240–245
Vitronectin, 240–241, 243
    mast cell adhesion, 140–141, 143

Z

Zaprinast, 195, 196